Praise for *It Makes Sense!* . . .

This is the only resource I've seen that actually addresses the need for primary students to learn to navigate a number path prior to the more abstract number line as recommended by NCTM. Students need the experiences of discrete sets of numbers before they move to the more abstract number line; through this resource, students are immersed in engaging lessons and games to learn how to make sense out of moving up and down a path, then move on to number lines when ready. There are opportunities for students to reason, make sense of the math, and communicate their understanding, all embedded into the activities. A must-have for all math teachers of young learners!

—Lisa Rogers, Director of Professional Learning Services, Math Solutions

We've known for years that great teaching, and even greater learning, requires touching, manipulating, drawing, constructing, representing, and making sense of the math being encountered in these ways. In this wonderfully teacher-friendly *It Makes Sense!*, Ann Carlyle captures the power of number paths and the number line to model practical ways of strengthening teaching and deepening learning of key primary grades' number skills and concepts.

—Steven Leinwand, American Institutes for Research

I have learned more about teaching mathematics from author and educator Ann Carlyle than anyone else, and have had the honor of learning from, being inspired by, and working with her for decades. In this resource, Ann demonstrates what an invaluable tool the number line is for elementary students learning math. Number lines promote accuracy, flexibility, and critical thinking as students make sense of the math problems they solve. Ann's friendly way of explaining, connecting, and modeling makes *It Makes Sense!* a tremendous resource for educators.

—Abbie Winter, Elementary Math Consultant and Coach, New Mexico

Ann Carlyle offers a valuable, much needed, and easy-to-use resource that masterfully helps young children navigate our number system and begin to develop base ten number concepts. Ann has a wealth of experience, and this resource is a testament to her knowledge and expertise as a teacher and teacher leader.

—Rusty Bresser, Lecturer and Supervisor of Teacher Education, University of California at San Diego

Finally! A resource that highlights the importance of early number lines and introduces early elementary educators to number paths. NCTM has touted the importance of number paths for decades. Ann has created a resource that is powerful, practical, and user-friendly. The lessons are rich, fresh, appropriate for young students, and conceptually based.

—TJ Jemison, International Elementary Math Consultant

Number lines are a powerful tool for fostering number sense. They encourage students to use benchmark numbers flexibly and they help illustrate the relationship between addition and subtraction. This is an invaluable resource, providing teachers with a cohesive set of lessons and games to bolster their students' number sense through the use of number paths and number lines.

—Amy Mayfield, Instructional Designer, Math Solutions

It Makes Sense!

Using Number Paths and Number Lines to Build Number Sense

Grades K–2

Ann Carlyle

Series Editor, Melissa Conklin

Math Solutions
Boston, Massachusetts, USA

Math Solutions, a division of Houghton Mifflin Harcourt Publishing Company
www.mathsolutions.com

Cataloging-in-Publication Data on file with the Library of Congress.

Math Solutions is a division of Houghton Mifflin Harcourt.

ISBN-13: 978-1-935099-51-2

Executive Editor: Jamie Ann Cross
Production Manager: Denise A. Botelho
Editorial Assistant: Kirby Sandmeyer
Cover Design: Susan Barclay, Barclay Design
Cover Image: Nell Campbell
Author Photo: Katie Bowman
Interior Design and Composition: MPS
Photo Credits: page 316, More/Less hand images © Dem10/Getty Images/Istockphoto.com and HOUGHTON MIFFLIN HARCOURT; Shopping Cart icon on back cover made by Freepik from www.flaticon.com

Printed in the United States of America.
1 2 3 4 5 6 7 8 9 10 0304 27 26 25 24 23 22 21 20
4510006959 ABCDE

A Message from Math Solutions

We at Math Solutions believe that teaching math well calls for increasing our understanding of the math we teach, seeking deeper insights into how students learn mathematics, and refining our lessons to best promote students' learning.

Math Solutions shares classroom-tested lessons and teaching expertise from our faculty of professional learning consultants as well as from other respected math educators. Our publications are part of the nationwide effort we've made since 1984 that now includes:

- more than five hundred face-to-face professional learning programs each year for teachers and administrators in districts across the country;
- professional learning books that span all math topics taught in kindergarten through high school;
- videos for teachers and for parents that show math lessons taught in actual classrooms;
- on-site visits to schools to help refine teaching strategies and assess student learning; and
- free online support, including grade-level lessons, book reviews, inservice information, and district feedback, all in our Math Solutions Online Newsletter.

For information about all of the products and services we have available, please visit our website at *www.mathsolutions.com*. You can also contact us to discuss math professional learning needs by calling (877) 234-7323 or by sending an email to *info@mathsolutions.com*.

We're always eager for your feedback and interested in learning about your particular needs. We look forward to hearing from you.

Math Solutions.
FOUNDED BY MARILYN BURNS
From Houghton Mifflin Harcourt.

To Tom, who knows the best way to get from here to there

Contents

(continued)

Second Grade

Section II: Lessons Using Number Paths and Number Lines

Section III: Games Using Number Paths and Number Lines 163

Reproducibles and Other Downloadable Materials

Downloadable materials that accompany this resource can be accessed by registering your book at www.mathsolutions.com/myonlineresources using the key code IMSNP. These materials include reproducibles and connections to standards. To register your book, please follow these steps.

How to Access Online Materials

1. Go to mathsolutions.com/myonlineresources and log in if you already have an account. If you do not have an account, click or tap the Create New Account button at the bottom of the Log In form.
2. Create an account, even if you have created one with the Math Solutions bookstore. You will receive a confirmation email when your account has been created.
3. Once your account has been created, you will be taken to the Product Registration page. Click Register on the product you would like to access (in this case, *It Makes Sense! Using Number Paths and Number Lines to Build Number Sense*).
4. Enter key code **IMSNP** and click or tap the Submit Key Code button.
5. Click or tap the Complete Registration button.
6. To access the materials at any time, visit your account page.

Key Code
IMSNP

Reproducibles

In addition to being available in this resource, the reproducibles can be downloaded from www.mathsolutions.com/myonlineresources by registering your resource (see the previous page for details) using key code IMSNP.

(continued)

Foreword

Calling *all* primary teachers! *It Makes Sense! Using Number Paths and Number Lines to Build Number Sense* is the resource you need to deepen your students' number sense. It is packed full of engaging lessons, powerful number talks, and enjoyable games designed to develop flexible thinking and a better understanding of number. Additionally, the directions are clear and concise, encouraging both teaching and learning success. In the fast-paced world we live in, we often want quick, meaningful activities that can be implemented without spending too much time reading or searching. Ann Carlyle, in step with the series' friendly and accessible format, offers just that!

I recall walking into my second-grade classroom the first year I taught (almost 20 years ago) and being handed the school's math textbook alongside several "math tools," including a ten-frame, a number line, and a hundreds chart. I eagerly hung the ten-frame and number line up on the wall, next to my brightly displayed alphabet, and stapled the hundreds chart near my calendar. The textbook had two lessons on using ten-frames, which I taught soon thereafter, checking off the preverbal box. The number line and hundreds chart, on the other hand, seemed invisible on the wall where they'd been posted, garnering little if any attention.

Thankfully, my math coach and Math Solutions professional learning consultants changed that. Under their mentorship, I learned how to integrate important tools like number lines and hundreds charts into my math lessons to best help students become flexible and fluent problem solvers.

Gradually I transformed each tool from "posters" on the classroom walls to hands-on, interactive learning tools enthusiastically used by both my students and me.

The evolution of the It Makes Sense! series has been a way for me to share the pedagogy, ideas, and activities I've learned over the years to maximize the use of powerful math tools. I encourage you to add this latest title to your ever-growing math tools library (the first two books being *It Makes Sense! Using Ten-Frames to Build Number Sense* and *It Makes Sense! Using the Hundreds Chart to Build Number Sense).* The It Makes Sense! series has received many accolades over the year, including being recognized as an AEP Distinguished Achievement award winner. As the author of the first two books and series editor of this third one, I continue to embrace the simplistic yet informative nature of the teaching directions and easy-to-implement lessons, routines, and games. And there's more in store; I encourage you to register your resource so that you have access to the online features Math Solutions offers as well as stay informed as more books come out.

My hope is that *It Makes Sense! Using Number Paths and Number Lines to Build Number Sense* and the other resources in this series inspire your teaching and increase your students' number sense.

—Melissa Conklin, Series Editor

Acknowledgments

I want to thank and acknowledge the people who helped me put this book together. Melissa Conklin was an active and tireless editor who offered suggestions for every lesson and game in this book. Her work on two previous books—*It Makes Sense! Using Ten Frames to Build Number Sense, Grades K–2* and *It Makes Sense! Using the Hundreds Chart to Build Number Sense, Grades K–2*—was invaluable in helping me find ways to organize my thoughts and teaching suggestions.

I appreciate all the help the Math Solution folks gave me: Jamie Cross for getting the manuscript going, Denise Botelho who did so very much to produce and design this collection of words and numbers, and Kirby Sandmeyer for her additional help with editing. They were all able to take sketches and scratches and make them reader friendly.

Two talented and exceptional teachers, Laura Calene and Jill Means, at Ellwood Elementary School in the Goleta Union School District opened up their classrooms to me and the teacher candidates I supervise from the University of California, Santa Barbara. The kindergarten and second-grade students in these classes were very eager to try the games and activities we posed. These student voices are found in many of the vignettes that illustrate this work with number lines and number paths.

And finally, thanks to the many people of the mathematics education community who made me think and inspired me over the years, including Marilyn Burns, Phil Daro, Julian Weissglass, Nicolas Branca, Bill Jacob, Sherry Parrish, Cathy Humphreys, Ruth Parker, and Jo Boaler.

How to Use This Resource

What is a number line?

A number line is typically a horizontal line with sequential numbers read left to right.

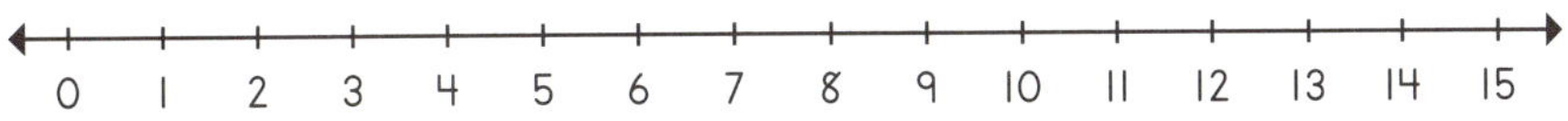

A number line.

"The arrangement of left to right is part of western culture, displayed on rulers, calendars, library bookshelves, floor signals above elevator doors, computer keyboards, and so on. The internalization of this convention starts in childhood: Young American children already explore sets of objects from left to right . . ." (Dehaene 1997, 82). The system of beginning at the left and moving to the right transfers to counting; students' mental representations of numbers becomes greater as they move to the right.

Teachers often have number lines located at the front of their classrooms alongside their alphabet display. Many primary school teachers construct a number path one day at a time, beginning on the first day of school and eventually building a number line that grows to or beyond one hundred. Multiples of tens are usually highlighted, and children count orally by ones, tens, and fives routinely while they or the teacher point to the numbers in order. In the primary grades, students use this tool to count forward and backward for addition and subtraction.

What is a number path?

In this book, the beginning lessons for young children use the structure of number paths to make the transition to the number line tool. On a number path, numbers are represented by rectangles, and each rectangle can be counted.

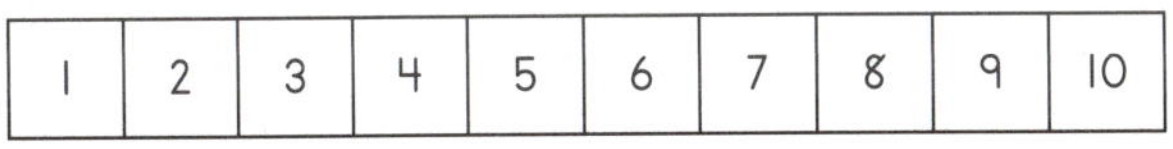

A number path.

The number line with structured units is a "measurement model." This means the numbers are representations of lengths, rather than labels for points on a line. The spaces between the numbers may be confusing to young children because the numbers stand for the progressive distance, a measure. Young children are much more comfortable with counting things, so a *number path*, which is a "counting model," often makes more sense to them.

What is an open number path and number line?

In this book, the lessons begin with putting number cards in order, and then move to using this constructed number path and thinking about how it can be used to help solve problems or play games. The lessons and games then transition to an open (sometimes referred to as *empty*) number path using blank color paper squares. The colors, in groups of five, help establish landmark numbers of 5, 10, 15, 20, and so forth. Many of the games and lessons in this book use these developmental stages to build meaning for working with an open (or empty) number line where students keep track of adding and subtracting along the line by marking their jumps forward or backward.

Instead of being a measurement model, the open number line is a "counting model." The marks on the line are not necessarily proportional. The lengths or distances between the marks do not necessarily indicate their value. The open number line is a visual representation for recording and sharing thinking strategies during the actual process of mental calculation.

One example of the open number line (showing 8 + 5) is as follows:

Example 1: An open (or empty) number line showing *8 + 5*.

In this example, the student decomposed the 5 to become 2 and 3. She then combined the 2 with the 8 to get 10 and then added three more to get 13. We might presume from this example that this student is comfortable with making sums to ten and chose to do the addition example this way instead of adding on five more ones to eight. When students reference the open number line, they make choices that use the number ideas they already know and understand. A different student, who may be more comfortable with doubles, might decompose the 8 in 8 + 5 into 5 and 3, and then add the two 5s to make 10 and then add three more to get 13.

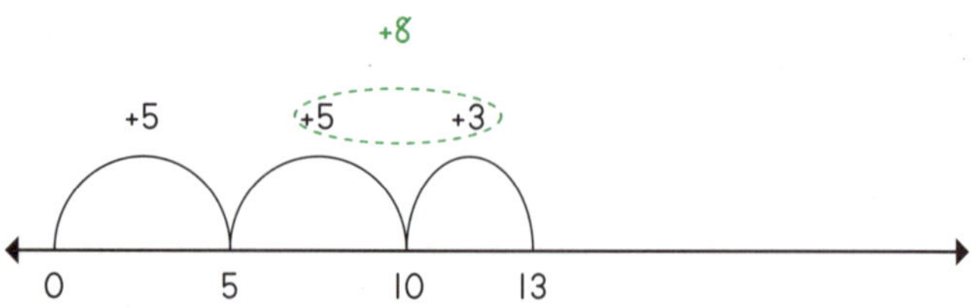

Example 2: An open (or empty) number line showing *8 + 5*.

The use of an open number line provides a window into how the students in Examples 1 and 2 are thinking about joining the two numbers 8 and 5. When students describe different ways of thinking about the same problem, they are making sense of the numbers they are combining, instead of just counting on.

This resource bridges the traditional number line (with all the numbers included from left to right) to an open number line through an open number path—a visual that displays the structure of the counting number system using two colors of paper squares in groups of fives and tens.

In Figure 1, using the paper squares, each shaded group of five red squares alternates with a group of five yellow squares. The thirty-five squares shown can be thought of as seven groups of five, or as three groups of ten with five more at the end of the line. Five is an important number to use because we have five fingers on each one of our two hands, and our number system is a base ten system.

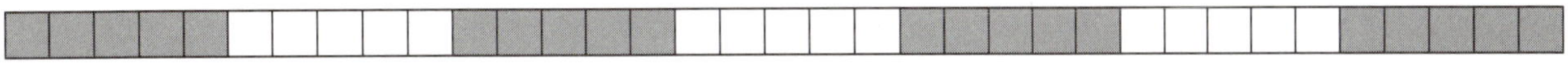

Figure 1: Moving to an open number line using paper squares (an open number path).

The magnitude of the number is represented by the location of paper squares on the line, rather than by the numbers used. This kind of visual is based on work with the arithmetic rack, or *rekenreks* (a Dutch word for arithmetic rack), that comes to us from The Freudenthal Institute for Science and Mathematics Education in the Netherlands. The structure offers visual support, and children can count five squares at a time, much like they do when using beads on rekenreks. We can apply the idea of beads on rekenreks to another useful model of an open number path. For this model, make a length of beads in two colors. This can help make the transition from rekenreks to the paper open number path.

Figure 2: Moving to an open number path using beads on a rekenrek.

There are many ways you will return to benchmark (friendly) numbers in the lessons and games introduced in this book. **Note:** Any combination of two colors can be used, just make sure there is a strong visual contrast.

Why Use Number Lines?

The open number line was originally proposed as a central model for addition and subtraction by researchers from the Netherlands in the 1980s. It was developed out of a need for a new tool to overcome problems experienced by children when performing addition and subtraction involving two-digit numbers.

There persists a common error, for example, when solving a problem such as *53 – 26* using the traditional vertical algorithm that involves regrouping. Children take the smaller number away from the larger one, and thus take 3 from 6. When using the open number line to subtract 26, the child would begin at 53 and jump back 20 (or two jumps of ten) to 33, and then jump back another six, or maybe two jumps of three, to arrive at 27. When using the open number line, the child thinks of the first number as a whole (53) instead of splitting it into tens and ones, as in the traditional paper-and-pencil algorithm.

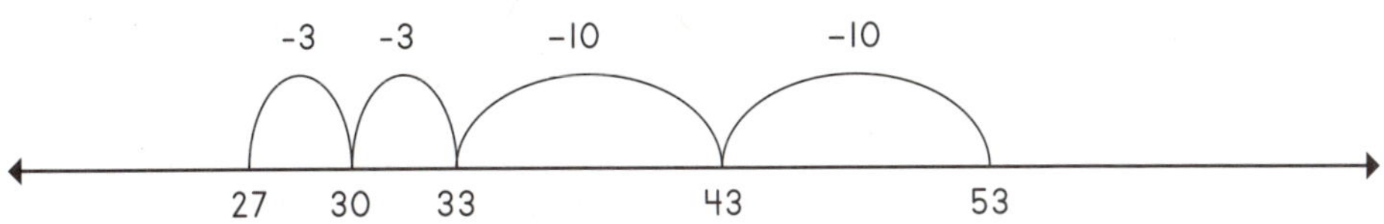

An open number line showing *53 – 26*.

In the upper grades, students are often introduced to a number line with negative and positive numbers, and they use this tool to learn how to add and subtract integers.

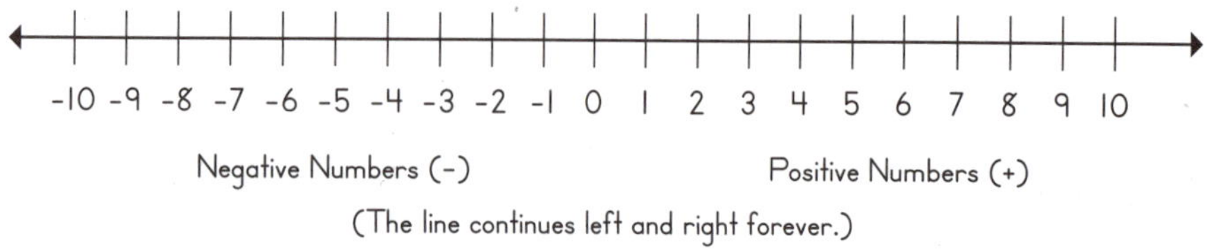

A number line with negative and positive numbers.

Sometimes, teachers use a vertical number line to represent the thermometer when they teach students about integers. Also, young children are often familiar with a vertical number line (based on inches) that is used to measure their height on a wall.

However, the number line can be limited as a model. Fractions and irrational numbers such as π or $\sqrt{2}$ are not easily displayed on a number line. But, with more exposure to number lines as visual and logical tools in the early grades, students are later able to make much more sense of fractions as numbers on a number line instead of thinking of fractions only as pieces of a pizza or parts of a circle.

Why Use These Lessons?

The introductory ideas, the lessons, and the games in *It Makes Sense! Using Number Paths and Number Lines to Build Number Sense* promote the flexible use of number lines to solve problems and scaffold or document students' thinking.

The open number line drawings that students make in these lessons become much more than procedures or algorithms. These drawings become windows into understanding how students construct their number ideas, and thus are valuable formative assessment tools.

How Do I Use the Lessons?

Three Categories

This resource is divided into three categories: introductory ideas, lessons, and games. The first section of introductory ideas provides directions for teachers to engage students in conversations about the structure and use of number paths and number lines. These minilessons are a great way to begin math class and support the development of student thinking—specifically, how to locate a range of numbers and compare these numbers to each other.

The lessons section features activities that introduce students to number paths and number lines as tools for thinking about numbers, their magnitude, and their relationship to each other. The lessons also encourage the use of these tools as frameworks for thinking about computation of one-, two-, and three-digit numbers.

The games section offers games that help students develop numerical fluency, as well as challenge them to think strategically and play cooperatively. The games section provides excellent opportunities for students to reinforce and practice the skills introduced in the lessons.

Where to Start

You can begin by using the ideas in Section I to support students in developing familiarity with the various number line models in kindergarten and first and second grade. You can then introduce the number line using the lessons in Section II.

In the "Related Lessons" sections of each introductory idea, lesson, or game, you get suggestions for what to teach next. Also, assessment opportunities are highlighted throughout the book to allow you to obtain formative assessment information about which lessons and games are appropriate for students to do next.

Lesson and Game Overview

In Sections II and III, each lesson and game opens with an overview that gives you an opportunity to become acquainted with the corresponding mathematical goals, as well as what students will be doing.

Time

The "Time" section in each lesson and game gives a general prediction of the time it takes to carry out the lesson. In general, each lesson takes thirty to forty minutes. Games usually take less time (ten to thirty minutes) to complete. Repetition is encouraged for many of the lessons and games. A "Time" section is not included in Section I only because a key characteristic of these minilessons is that they take but a brief amount of time, typically five to fifteen minutes.

Materials

The following is a basic list of materials needed. In Sections II and III, each lesson and game begins with a specific list.

- whiteboard (or poster paper) and markers
- sticky notes
- counters (such as color tiles, interlocking cubes, lima beans, or pennies)
- 3-inch-by-3-inch paper squares arranged in groups of five using two contrasting colors such as yellow and red (used to create number paths)
- dice
- clothespins
- paper lunch sacks
- children's books (to be read as an introduction to the topics and content in many of the lessons and games)
- reproducibles (Whenever possible, reproducibles are included to facilitate the implementation of the suggested ideas and activities; see page x to learn how to access reproducibles in a downloadable, printable format.)

Key Questions

Each introductory idea, lesson, and game offers key questions to promote student thinking, class discussion, and assessment of what students know. These carefully planned questions elicit deeper thinking and reasoning among students and are meant to be asked throughout each lesson. Often, it is necessary to scribe or record student thinking. Recording student thinking connects a child's thinking to representations (such as drawing jumps on a number line) or to symbols (such as numbers). It allows the student who is speaking, and others in the class, to observe their thinking visually.

Teaching Directions

The teaching directions are presented in a step-by-step lesson plan with references to when and how to use the key questions, and what a student might be thinking. Some of the lessons are divided into parts to make the planning process more manageable.

Teacher Reflections

Descriptions of what took place in the classroom help you think about teaching strategies and support you in thinking about your own students. These lessons and games have been tested with real children in regular classrooms, and they are student-tested favorites. The reflections are written to help you see how children think, reason, and communicate about numbers.

Additional Teaching Insights

In addition to what has been shared so far, teaching insights are provided throughout the lessons so you can refer to them easily time and time again. The insights are categorized as follows:

Assessment Opportunities

Highlighted formative assessment ideas support you in learning more about the mathematical thinking your students are using when solving problems or playing games.

A Child's Mind . . .

These insights inform you about common student errors and/or misconceptions, so you can teach more purposefully. Teachers need to think about what goes on in a child's mind and then find ways to address how the child reasons.

Differentiating Your Instruction

Many suggestions are provided so you can meet the needs of a wide range of students. Sometimes the ideas involve making the range of the numbers larger or smaller; other times, the structure of the problems or activities can be made more or less complex.

Literature Connection

Children's literature brings another voice, or part of the outside world, into classrooms. Because children are naturally drawn to books and stories, ideas for including these resources are highlighted.

Math Matters!

Mathematical definitions, research summaries, learning scenarios, and math content are described to support math learning in K–2 classrooms. These sections provide an opportunity to deepen one's own math content knowledge.

Math Talk

At the heart of this resource is an emphasis on promoting and supporting student *math talk*—thinking and reasoning—in various settings. Strategies such as Think-Pair-Share are explained and incorporated into lessons and activities.

Teaching Tip

Teaching tips highlight information that is especially helpful to you in managing your classroom and implementing the activities and lessons successfully. Explanations and examples of student responses give you a glimpse of how lessons play out.

Technology Tip

These tips offer insights for using technology to make a lesson run more smoothly. For example, using a whiteboard to display and interact with a number path or number line.

Time Saver

Teachers are always looking for ways to find more time in the teaching day for instruction. Many of these time-saving ideas involve using previously created number paths or student helpers who take on more responsibility in preparing and/or teaching the lesson.

Connections to Standards

The ideas in this book can be used alongside your own curriculum, standards, or pacing guides. Specific connections to standards are available online. To access these connections, follow the instructions found on page x.

More Resources in the Series!

We are excited to share this resource, *It Makes Sense! Using Number Paths and Number Lines to Build Number Sense, K–2,* as one of several resources in the It Makes Sense! series. For information on these titles and others, see www.mathsolutions.com.

Introductory Ideas, Lessons, and Games by Grade

The introductory ideas, lessons, and games in this resource can be used for grades K–2, depending on the ability of students. This table reflects my suggestions.

Title	Recommended Grade Level			Page
	K	1	2	
I-1 Using an Open Number Path without Numerals 1–20	X			5
I-2 Using a Number Path with Numerals 1–10	X			9
I-3 Using a Number Path with Numerals 1–20	X			12
I-4 Using a Number Path with Numerals 1–30	X			16
I-5 Using an Open Number Path without Numerals 1–60		X		20
I-6 Using a Number Path with Numerals 1–30		X		24
I-7 Using a Number Path with Numerals 1–50		X		29
I-8 Using a Number Path with Numerals 31–60		X		33
I-9 Using a Number Path with Numerals 51–100		X		37
I-10 Using an Open Number Line with Numerals 0 and 50		X		41
I-11 Using an Open Number Line with Numerals 0 and 100		X		45
I-12 Using an Open Number Path without Numerals 1–60			X	50

(continued)

Title	Recommended Grade Level			Page
	K	1	2	
G-6 101 and Out!		X	X	215
G-7 Hot Lava Bridge: Forward and Back, More and Less		X	X	225
G-8 The Larger Difference Game		X	X	234
G-9 Adding Nines, Tens, and Elevens		X	X	247
G-10 The Smaller Difference Game		X	X	261
G-11 Get to the Target (Version 1)		X		270
G-12 Get to the Target (Version 2)		X	X	277
G-13 Get to the Target (Version 3)			X	285
G-14 Race to 1,000			X	293

Introducing Number Paths and Number Lines

Kindergarten

First Grade

(*Continued*)

Second Grade

Why this section?

This section offers sixteen minilessons for introducing number paths and number lines to students. These introductory ideas should be used, as appropriate, prior to any of the lessons and games in the subsequent sections.

These minilessons can also be used as a formative assessment tool that allows you to target instruction around ideas that may be confusing to a small group of students. To do so, make note of student interest, when students become confused or distracted, or when students are able to articulate their thinking and give reasons for the procedures they choose to use. Last but not least, these minilessons can be used to share math topics with parents, encouraging exploration at home.

These ideas encourage students to share thinking. When students explain their thinking, the classroom becomes a place where the source of the strategies and information comes from the students instead of the teacher. When students listen to their peers, they often are able to develop their own more accurate, efficient, or flexible strategies for thinking.

How much time do these minilessons take?

These ideas shouldn't take more than ten minutes or so (hence "minilessons"). Young children are often not comfortable sitting for longer periods than this. I recommend that you prepare a collection of these minilessons that are ready to use when the need arises. You can shorten the amount of time of each by limiting the number of questions, if necessary. However, make sure to give students enough time to explain and support their thinking.

What is your role as teacher during these minilessons?

It is important for the classroom environment to be a safe place, where students are comfortable sharing their thinking. This means all answers are accepted and students listen to one another, feel comfortable agreeing or disagreeing with one another, and generally understand that the purpose of their conversations is to make sense of the content and learn how to explain their thinking. If students seem confused, change the question or ask students to share what they notice. It is important for children to be able to come up and touch the number path or line to explain their thinking if they want. Strategies such as think-pair-share are useful for students to practice expressing their ideas before speaking to the entire class. It takes time to develop this kind of learning community, but it is well worth it. Strive for a light touch while listening carefully to student thinking.

Using an Open Number Path without Numerals 1–20

Introducing 1

Related Lessons and Games

- L-1 Building a Number Path
- G-1 Mystery Number (Version 1)

Key Questions

- How did you know there were ten?
- Do you think there are the same number of red squares as yellow squares? How can you find out?
- How can you tell how many red and yellow squares there are without counting them one by one?
- How do you know that red squares come next?

Suggested Grade Level

Kindergarten

Materials

open number path, created using red and yellow paper squares, ten of each color, organized in groups of five starting with red (See Reproducible A, How to Make an Open Number Path)

Teaching Directions

1. Prepare an open number path per the materials list. Make sure the number path is large enough for everyone in class to see it.

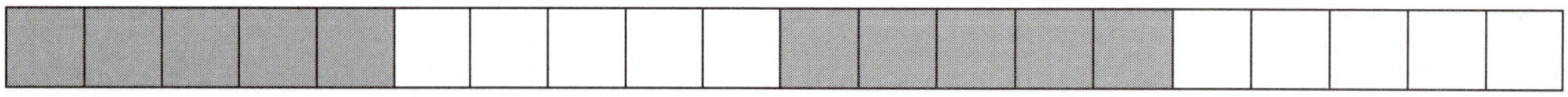

Open number path.

2. Gather students on the rug so everyone has a good view of the number path. Seat them close together to make listening to each other easier.
3. Draw students' attention to the number path. Say, "This is a number path. I made this number path with some red and yellow squares." Ask students to look at the path and think about the question: How many red squares are there?
4. Remind students to keep their answer in their head while everyone gets time to think. Ask them to give you a quiet signal, such as a thumbs-up on their chest, when they have an answer to share.
5. Now share with students, "I am going to count to three. When I say 'three,' I want everyone to whisper the number of red squares that you see." You may hear students respond with "ten" or "five."
6. Call on a student to explain their thinking. Ask, "How did you know there were ten?" Most students will likely come up and count each square, touching them with their fingers.

A Child's Mind . . .

Providing Time to Think

Asking students to "whisper" their answer is a helpful way to give all students an opportunity to "say the answer." It also de-emphasizes the significance of saying the answer and allows students to focus more on reasoning and strategies.

7. If a student thinks there are five red squares, ask, "What if I wanted to know about *all* the red squares on the path? Does someone have another way of knowing how many red squares there are on this path?" Some students may have noticed that there are two groups of five red squares and may say that five and five are ten.

8. Continue, "Now I'd like you to think about how many yellow squares there are on the path." Remind students to keep their answer in their head while everyone has time to think. Ask them to give you a quiet signal, such as a thumbs-up on their chest, when they have an answer to share.

9. Next, count out loud to three and repeat the questioning process, giving different students an opportunity to share their thinking. Encourage them to think about how the path is made. Mimic counting out five red squares, gluing them down, then counting out five yellow squares and gluing them down, then five more red squares, and then five more yellow squares.

10. It is helpful if students notice there are the same number of red squares as yellow squares. If this does not come up, ask, "Do you think there are the same number of red squares as yellow squares? How can you find that out?"

11. Now, fold the number path in half so that students can only see ten squares: five red and five yellow. Have students watch as you do this.

12. Point to half of the number path and say, "Think to yourself, 'How many red and yellow squares are in this part?' Give me a quiet signal when you have your answer."

A Child's Mind . . .

Touching the Number Path

When reading, children start out by touching each word as they read it. Later, they begin to move their finger under a line of text as they read more fluently. So, too, in math do children begin by touching each number, one at a time, as they count on a number path or number line. As their thinking develops, they become more comfortable counting in their head.

Math Talk

Think-Pair-Share

When teachers ask a question to the whole class, often one student responds. However, when teachers use the think-pair-share strategy, all students get a chance to talk about the question. This can be an opportunity for students to rehearse with their partner what they may want to say to the whole class, or a chance for them to predict what will happen in a safe setting. With experience, students become comfortable with the think-pair-share strategy and appreciate the opportunity to express their ideas before voicing them to the whole class.

Discussion Buddies

Often, students don't know who to turn to when teachers ask them to talk to a partner. To save time and alleviate any issues that might arise, instruct students to share their thinking specifically with a *shoulder partner*.

13. Ask students to whisper their answers after you count to three. Students are likely to respond with "Five red and five yellow" or "Ten red and yellow squares."
14. Ask, "How can we tell how many red and yellow squares there are in this part without counting them one by one?" Use the think-pair-share strategy to bolster students' willingness to share and give everyone the opportunity to share. To do this, tell students to think alone first. Then direct them to share their thinking with a shoulder partner. Next, encourage someone to share with the whole class. Promote active listening by asking another student to repeat what they heard the first student share. Then ask, "Who has a different way to tell how many squares are in this part without counting them one by one?" Encourage students to share their different strategies.
15. Now unfold the path so students see all twenty squares again. Remind students, "Here we have our twenty squares. Ten are red and ten are yellow. What if we wanted to make this path longer and glue on more squares? What color should we use first?"
16. Ask for an explanation, "How do you know that red squares come next?"
17. Bring the discussion to a close by asking, "How many squares would we then have in all if we added five more red squares?" This question challenges students to think abstractly because they are being asked for the total number if five more were to be added to twenty, but the physical representation has not been altered.

Teaching Tip

Keep the Number Path on Display

It is helpful to keep the number path on display in an accessible place so children can go to it and count the squares again and again. Adding on to the original twenty squares is a project that some students find very engaging.

Using a Number Path with Numerals 1–10

Introducing

2

Related Lesson

- L-2 Counting Cups on the Number Path

Key Questions

- How did you know which number is (numbers are) hiding?
- What is another way to know that [number] is behind the sticky note?
- How can we tell which number is more and which number is less?

Suggested Grade Level

Kindergarten

Materials

number path labeled *1* through *10*, with the numbers spaced equally apart (approximately 3 inches), created using paper sentence strips or cash register paper that is approximately 3 inches tall

large sticky notes, approximately 3-inch-by-3-inch squares

Teaching Directions

1. Prepare a number path per the materials list. Make sure the number path is large enough for everyone in class to see it.

1	2	3	4	5	6	7	8	9	10

Number path labeled *1* through *10*.

2. Gather students on the rug so that everyone has a good view of the number path. Seat them close together to make listening to each other easier.
3. Draw students' attention to the number path. Say, "This is a number path. I am going to hide one of the numbers on the number path. I want you to figure out which number is hiding."
4. Ask students to look at the floor and cover their eyes. While they do this, cover one of the numbers using a sticky note.
5. Instruct students to look up and uncover their eyes. Ask, "Which number is hiding?" Remind students to keep their answer in their head while you give everyone time to think. Ask them to give you a quiet signal, such as a thumbs-up on their chest, when they have an answer to share.
6. Now say, "I am going to count to three. When I say 'three,' I want everyone to whisper the name of the number that is hiding."
7. Call on a student to explain his or her thinking. Ask, "How did you know which number is hiding?" Most students will likely determine the hidden number by noticing what it comes after. For example, "Eight is hiding because the sticky note is after seven and I know eight comes after seven." Encourage students to consider which number follows the hidden number by asking, "What is another way to know that eight is behind

A Child's Mind . . .

Providing Time to Think

Asking students to "whisper" their answer is a helpful way to give all students an opportunity to "say the answer." It also de-emphasizes the significance of saying the answer and allows students to focus more on reasoning and strategies.

the sticky note?" (Possible response: "It's just before nine.") Remove the sticky note.

8. Ask students to cover their eyes again while you hide a different number. Repeat the process, giving different students an opportunity to share their thinking.
9. Repeat the process one more time, but this time use two sticky notes to hide two different numbers at the same time. (They can be two consecutive numbers or two numbers that are not consecutive.)
10. Count to three again and ask students to whisper the name of the two numbers that are hiding. Then ask, "How do you know?" and call on several students to share their reasoning. Before moving on, place the two sticky notes so they are below the numbers they covered (so the numbers are visible). The sticky notes serve as a reminder of the two numbers that were hidden.
11. Now ask, "How can we tell which number is less and which number is more?" Use the think-pair-share strategy to bolster students' willingness to share and give everyone the opportunity to share. To do this, tell students to think alone first. Then direct them to share their thinking with a shoulder partner. Next, encourage someone to share with the whole class. Promote active listening by asking another student to repeat what they heard the first student share. Then ask, "Who has a different way to tell which number is more?" Encourage students to share their different strategies.
12. Close the discussion by making the number path and some sticky notes available to students to explore and investigate on their own during independent class times. (This activity could be available as a learning station during a math workshop.) Students are likely to use more than two or three sticky notes to challenge each other about the hidden numbers.

Math Talk

Think-Pair-Share
When teachers ask a question to the whole class, often one student responds. However, when teachers use the think-pair-share strategy, all students get a chance to talk about the question. This is an opportunity for students to rehearse with a partner what they may want to say to the whole class, or a chance for them to predict what will happen in a safe setting. With experience, students become comfortable with the think-pair-share strategy and appreciate the opportunity to express their ideas before voicing them to the whole class.

Discussion Buddies
Often, students don't know who to turn to when teachers ask them to talk to a partner. To save time and alleviate any issues that might arise, instruct students to share their thinking specifically with a *shoulder partner*.

A Child's Mind . . .

Talking about *More* and *Less*
Young children are typically more confident in talking about the idea of "more" than they are with "less." Perhaps this is because they are so interested in having "more" than "less" of treats, toys, or attention. With young learners, it helps to use gestures such as hands wide apart for "more" and hands close together for "less."

Introducing 3

Using a Number Path with Numerals 1–20

Suggested Grade Level

Kindergarten

Materials

number path labeled *1* through *20*, with the numbers spaced equally apart (approximately 3 inches), created using paper sentence strips or cash register paper that is approximately 3 inches tall

large sticky notes, approximately 3-inch-by-3-inch squares

Related Lesson

- L-2 Counting Cups on the Number Path

Key Questions

- How did you know which number is (numbers are) hiding?
- What is another way to know that [number] is behind the sticky note?
- How can we tell which number is more and which number is less?

Teaching Directions

1. Prepare a number path per the materials list. Make sure the number path is large enough for everyone in class to see it.

1	2	3	4	5	6	7	8	9	10	11	12	13	14	15	16	17	18	19	20

Number path labeled *1* through *20*.

2. Gather students on the rug so that everyone has a good view of the number path. Seat them close together to make listening to each other easier.
3. Draw students' attention to the number path. Say, "I am going to hide one of the numbers on the number path. I want you to figure out which number is hiding."
4. Ask students to look at the floor and cover their eyes. While they are doing this, cover one of the numbers using a sticky note.
5. Instruct students to look up and uncover their eyes. Ask, "Which number is hiding?" Remind students to keep their answer in their head while you give everyone time to think. Ask them to give you a quiet signal, such as a thumbs-up on their chest, when they have an answer to share.
6. Now say, "I am going to count to three. When I say 'three,' I want everyone to whisper the name of the number that is hiding."
7. Call on a student to explain their thinking. Ask, "How did you know which number is hiding?" Most students will likely determine the hidden number by noticing what it comes after. For example, "Eighteen is hiding because the sticky note is after seventeen and I know eighteen comes after seventeen." Encourage students to consider which number follows the hidden number by asking, "What is another way to know

A Child's Mind . . .

Providing Time to Think
Asking students to "whisper" their answer is a helpful way to give all students an opportunity to "say the answer." It also de-emphasizes the significance of saying the answer and allows students to focus more on reasoning and strategies.

that eighteen is behind the sticky note?" Possible responses: "It's just before nineteen" or "When you count, you say sixteen, seventeen, eighteen, nineteen, twenty." Remove the sticky note.

8. Ask students to cover their eyes again while you hide a different number. Repeat the process, giving different students an opportunity to share their thinking.

9. Repeat the process one more time, but this time use two sticky notes to hide two different numbers at the same time. (They can be two consecutive numbers or two numbers that are not consecutive.)

10. Count to three again and ask students to whisper the names of the two numbers that are hiding. Then ask, "How do you know?" and call on several students to share their reasoning. Before moving on, place the two sticky notes so they are below the numbers they covered (so the numbers are visible). The sticky notes serve as a reminder of the two numbers that were hidden.

11. Now ask, "How can we tell which number is less and which number is more?" Use the think-pair-share strategy to bolster students' willingness to share and give everyone the opportunity to share. To do this, tell students to think alone first. Then direct them to share their thinking with a shoulder partner. Next, encourage someone to share with the whole class. Promote active listening by asking another student to repeat what they heard the first student share. Then ask, "Who has a different way to tell which number is more?" Encourage students to share their different strategies.

Math Talk

Think-Pair-Share

When teachers ask a question to the whole class, often one student responds. However, when teachers use the think-pair-share strategy, all students get a chance to talk about the question. This is an opportunity for students to rehearse with a partner what they may want to say to the whole class, or a chance for them to predict what will happen in a safe setting. With experience, students become comfortable with the think-pair-share strategy and appreciate the opportunity to express their ideas before voicing them to the whole class.

Discussion Buddies

Often, students don't know who to turn to when teachers ask them to talk to a partner. To save time and alleviate any issues that might arise, instruct students to share their thinking specifically with a *shoulder partner*.

A Child's Mind . . .

Talking about *More* and *Less*

Young children are typically more confident in talking about the idea of "more" than they are with "less." Perhaps this is because they are so interested in having "more" than "less" of treats, toys, or attention. With young learners, it helps to use gestures such as hands wide apart for "more" and hands close together for "less."

12. Now cut the number path in half and display the second half under the first half. Ask students to watch as you do this. Your number path should look something like this:

1	2	3	4	5	6	7	8	9	10
11	12	13	14	15	16	17	18	19	20

Number path labeled *1* through *20*, but it is now cut in half, with the second half below the first.

13. Ask, "What do you notice about the number paths now?" Possible responses may be that the second number in the numbers below is the same as the top number.
14. Ask, "What else do you notice about this arrangement?" Possible responses may be that each number in the bottom row has a one in it except for the last number.
15. Say, "Let's read each of these numbers in order as I point." Count out from one to twenty, pointing to each number.
16. Close the discussion by making the number path and some sticky notes available to students to explore and investigate on their own during independent class times. (This activity could be available as a learning station during math workshop.) Students are likely to use more than two or three sticky notes to challenge each other about the hidden numbers.

Introducing 4 Using a Number Path with Numerals 1–30

Suggested Grade Level

Kindergarten

Materials

number path labeled *1* through *30*, with the numbers spaced equally apart (approximately 3 inches), created using paper sentence strips or cash register paper that is approximately 3 inches tall

large sticky notes, approximately 3-inch-by-3-inch squares

Related Lesson

- L-2 Counting Cups on the Number Path

Key Questions

- How did you know which number is (numbers are) hiding?
- What is another way to know that [number] is behind the sticky note?
- How can we tell which number is more and which number is less?

Teaching Directions

1. Prepare a number path per the materials list. Make sure the number path is large enough for everyone in class to see it.

1	2	3	4	5	6	7	8	9	10	…	21	22	23	24	25	26	27	28	29	30

Number path labeled *1* through *30 (abbreviated here in order to fit on this page).*

2. Gather students on the rug so that everyone has a good view of the number path. Seat them close together to make listening to each other easier.
3. Draw students' attention to the number path. Say, "I am going to hide one of the numbers on the number path. I want you to figure out which number is hiding."
4. Ask students to look at the floor and cover their eyes. While they are doing this, cover one of the numbers using a sticky note.
5. Instruct the students to look up and uncover their eyes. Ask, "Which number is hiding?" Remind students to keep their answer in their head while you give everyone time to think. Ask them to give you a quiet signal, such as a thumbs-up on their chest, when they have an answer to share.
6. Now say, "I am going to count to three. When I say 'three,' I want everyone to whisper the name of the number that is hiding."
7. Call on a student to explain his or her thinking. Ask, "How did you know which number is hiding?" Most students will likely determine the hidden number by noticing what it comes after. For example, "Twenty-three is hiding because the sticky note is after twenty-two and I know twenty-three comes after twenty-two." Encourage students to consider which number follows the hidden number by asking, "What is

A Child's Mind . . .

Providing Time to Think
Asking students to "whisper" their answer is a helpful way to give all students an opportunity to "say the answer." It also de-emphasizes the significance of saying the answer and allows students to focus more on reasoning and strategies.

Math Talk

Think-Pair-Share

When teachers ask a question to the whole class, often one student responds. However, when teachers use the think-pair-share strategy, all students get a chance to talk about the question. This can be an opportunity for students to rehearse with a partner what they may want to say to the whole class, or a chance for them to predict what will happen in a safe setting. With experience, students become comfortable with the think-pair-share strategy and appreciate the opportunity to express their ideas before voicing them to the whole class.

Discussion Buddies

Often, students don't know who to turn to when teachers ask them to talk to a partner. To save time and alleviate any issues that might arise, instruct students to share their thinking specifically with a *shoulder partner*.

A Child's Mind . . .

Talking about *More* and *Less*

Young children are typically more confident in talking about the idea of "more" than they are with "less." Perhaps this is because they are so interested in having "more" than "less" of treats, toys, or attention. With young learners, it helps to use gestures such as hands wide apart for "more" and hands close together for "less."

another way to know that twenty-three is behind the sticky note?" Possible response: "It's just before twenty-four" or "When you count, you say twenty-one, twenty-two, twenty-three, twenty-four, twenty-five." Remove the sticky note.

8. Ask students to cover their eyes again while you hide a different number. Repeat the process, giving different students an opportunity to share their thinking.
9. Repeat the process one more time, but this time use two sticky notes to hide two different numbers at the same time. (They can be two consecutive numbers or two numbers that are not consecutive.)
10. Count to three again and ask students to whisper the names of the two numbers that are hiding. Then ask, "How do you know?" and call on several students to share their reasoning. Before moving on, place the sticky notes so they are below the two numbers they covered (so that the numbers are visible). The sticky notes serve as a reminder of the two numbers that were hidden.
11. Now ask, "How can we tell which number is less and which number is more?" Use the think-pair-share strategy to bolster students' willingness to share and give everyone the opportunity to share. To do this, tell students to think alone first. Then direct them to share their thinking with a shoulder partner. Next, encourage someone to share with the whole class. Promote active listening by asking another student to repeat what they heard the first student share. Then ask, "Who has a different way to tell which number is more?" Encourage students to share their different strategies.

12. Now cut the number path into thirds and display each third, one under the other. Ask students to watch as you do this. Your number path should look something like this:

1	2	3	4	5	6	7	8	9	10
11	12	13	14	15	16	17	18	19	20
21	22	23	24	25	26	27	28	29	30

Number path labeled *1* through *30*, but it is now cut in thirds, with the second third below the first, and the third part below the second.

13. Ask, "What do you notice about the number paths now?" Possible responses may be that the second number in the numbers below the top path is the same as the numbers above. Let several students come up to the display and point to the patterns they see in the arrangement.
14. Ask, "What else do you notice about this arrangement?" Possible responses may be that each number in the middle row has a one in it except for the last number and each number in the bottom row has a two in it except for the last number.
15. Say, "Let's read each of these numbers in order as I point." Count out from one to thirty, pointing to each number.
16. Close the discussion by making the number path and some sticky notes available to students to explore and investigate on their own during independent class times. (This activity could be available as a learning station during math workshop.) Students are likely to use more than two or three sticky notes to challenge each other about the hidden numbers.

A Child's Mind . . .

Counting Past Ten

It is typical for students, when counting past ten, to pause at nineteen and twenty-nine when they are counting as they mentally review the change in the pattern of the digits in the ones place.

Teaching Tip

Displaying a Hundreds Chart

If you have a pocket hundreds chart, this is an appropriate time to display this chart with numbers to thirty. Some of the numbers could be turned over so students could play a guessing game independently or in pairs.

Introducing

5

Using an Open Number Path without Numerals 1–60

Suggested Grade Level

Grade 1

Materials

open number path, created using red and yellow paper squares, 30 of each color, organized in groups of five starting with red (See Reproducible A, How to Make an Open Number Path)

Related Lesson

- L-3 Building the Open Number Path

Key Questions

- How did you know there were thirty?
- Do you think there are the same number of red squares as yellow squares? How can you figure this out?
- How can you tell how many red and yellow squares there are without counting them one by one?
- How do you know that red squares come next?

Teaching Directions

1. Prepare an open number path per the materials list. Make sure the number path is large enough for everyone in class to see it.

Open number path (abbreviated here in order to fit on this page).

2. Gather students on the rug so that everyone has a good view of the number path. Seat them close together to make listening to each other easier.
3. Draw students' attention to the number path. Say, "This is a number path. I made this number path with some red and yellow squares." Ask students to look at the path and think about the question: How many red squares are there?
4. Remind students to keep their answer in their head while everyone gets time to think. Ask them to give you a quiet signal, such as a thumbs-up on their chest, when they have an answer to share.
5. Now share with students, "I am going to count to three. When I say 'three,' I want everyone to whisper the number of red squares that you see." You may hear students respond with "thirty," counting six groups of five.
6. Call on a student to explain their thinking. Ask, "How did you know there were thirty?" Many students will likely come up and count each square, touching them with their fingers.
7. If a student thinks there are five red squares, ask, "What if I want to know about *all* the red squares on the path, not just one group of five?" Say, "Let's count by fives." Place your hands like parentheses around each group of five red squares as you count.

A Child's Mind . . .

Providing Time to Think
Asking students to "whisper" their answer is a helpful way to give all students an opportunity to "say the answer." It also de-emphasizes the significance of saying the answer and allows students to focus more on reasoning and strategies.

A Child's Mind . . .

Touching the Number Path

When reading, children start out by touching each word as they read it. Later, they begin to move their finger under a line of text as they read more fluently. So, too, in math do children begin by touching each number, one at a time, as they count on a number path or number line. As their thinking develops, they become more comfortable counting in their head.

8. Continue, "Now I'd like you to think about how many yellow squares there are on the path." Remind students again to keep their answer in their head while you give everyone time to think. Ask them to give you a quiet signal, such as a thumbs-up on their chest, when they have an answer to share.

9. Count again to three and repeat the process, giving different students an opportunity to share their thinking. Encourage students to think about how the path was made. Mimic counting out five red squares, gluing them down, then counting out five yellow squares and gluing them down, then five more red squares, then five more yellow squares, and so forth.

10. It is helpful if students notice there is the same number of red squares as yellow squares. If this does not come up, ask, "Do you think there is the same number of red squares as yellow squares? How can you figure that out?" Allow a couple of students to share their thinking.

11. Now ask, "How many red *and* yellow squares are there on this path? How do you know without counting each square?" One possible response is to count by tens, because each group of five red and five yellow squares equal ten squares. If this is suggested, say, "Let's count by tens." Then, proceed to do so.

12. Now fold the number path in half so that students can only see thirty squares (fifteen red and fifteen yellow). Have students watch as you do this.

13. Point to the folded number path and say, "Think to yourself: How many red and yellow squares are in this part? Give me a quiet signal when you have your answer."

14. Ask students to whisper their answers after you count to three. Students are likely to respond with "thirty" or "ten and ten and ten."

15. Ask, "How can you tell how many red and yellow squares there are in this part without counting them one by one?" Use the think-pair-share strategy to bolster students' willingness to share and give everyone the opportunity to share. To do this, tell students to think alone first. Then direct them to share their thinking with a shoulder partner. Next, encourage someone to share with the whole class. Promote active listening by asking another student to repeat what they heard the first student share. Then ask, "Who has a different way to tell how many squares are in this part without counting them one by one?" A possible response is "Thirty, because thirty and thirty is sixty." Some students might know this because they know three and three is six, so three tens and three tens make six tens.

16. Now unfold the path so students see all sixty squares again. Remind students, "Here we have our sixty squares. Thirty are red and thirty are yellow. What if we wanted to make this path longer and glue on more squares? What color should we use next?"

17. Ask for an explanation, "How do you know that red squares come next?"

18. Bring this discussion to a close by asking, "How many squares would we then have in all if we added five more red squares?" This question challenges students to think abstractly because they are being asked for the total number if five more were added to sixty, but the physical representation has not been altered.

Math Talk

Think-Pair-Share

When teachers ask a question to the whole class, often one student responds. However, when teachers use the think-pair-share strategy, all students get a chance to talk about the question. This can be an opportunity for students to rehearse with a partner what they may want to say to the whole class, or a chance for them to predict what will happen in a safe setting. With experience, students become comfortable with the think-pair-share strategy and appreciate the opportunity to express their ideas before voicing them to the whole class.

Discussion Buddies

Often, students don't know who to turn to when teachers ask them to talk to a partner. To save time and alleviate any issues that might arise, instruct students to share their thinking specifically with a *shoulder partner*.

Teaching Tip

Keeping the Number Path on Display

It is helpful to keep this path on display in an accessible place, so children can go to it and count the squares again and again. Adding on to the original sixty squares is a project that some students find very engaging. Making a path that is one hundred squares long is a nice goal for first grade.

Introducing 6

Using a Number Path with Numerals 1–30

Suggested Grade Level

Grade 1

Materials

number path labeled *1* through *30*, with the numbers spaced equally apart (approximately 3 inches), created using paper sentence strips or cash register paper that is approximately 3 inches tall

large sticky notes, approximately 3-inch-by-3-inch squares

Related Lesson

- L-4 Jumping by Ones and Tens

Key Questions

- How did you know which number is (numbers are) hiding?
- What is another way to know that [number] is behind the sticky note?
- How can we tell which number is less (the least) and which number is more (the most)?

Teaching Directions

1. Prepare a number path per the materials list. Make sure the number path is large enough for everyone in class to see it.

1	2	3	4	5	6	7	8	9	10	…	21	22	23	24	25	26	27	28	29	30

Number path labeled *1* through *30* (abbreviated here in order to fit on this page).

2. Gather students on the rug so that everyone has a good view of the number path. Seat them close together to make listening to each other easier.
3. Draw students' attention to the number path. Say, "This is a number path. I am going to hide one of the numbers on the number path. I want you to figure out which number is hiding."
4. Ask students to look at the floor and cover their eyes. While they are doing this, cover one of the numbers with a sticky note.
5. Instruct students to look up and uncover their eyes. Ask, "Which number is hiding?" Remind students to keep their answer in their head while you give everyone time to think. Ask them to give you a quiet signal, such as a thumbs-up on their chest, when they have an answer to share.
6. Now say, "I am going to count to three. When I say 'three,' I want everyone to whisper the name of the number that is hiding."
7. Call on a student to explain their thinking. Ask, "How did you know which number is hiding?" Most students will likely determine the hidden number by noticing what it comes after. For example, "Twenty-three is hiding because the sticky note is after twenty-two and I know twenty-three comes after twenty-two." Encourage students to

A Child's Mind . . .

Providing Time to Think

Asking students to "whisper" their answer is a helpful way to give all students an opportunity to "say the answer." It also de-emphasizes the significance of saying the answer and allows students to focus more on reasoning and strategies.

consider which number follows the hidden number by asking, "What is another way to know that twenty-three is behind the sticky note?" Possible response: "It's just before twenty-four" or "When you count, you say twenty-one, twenty-two, twenty-three, twenty-four, twenty-five." Remove the sticky note.

8. Ask students to cover their eyes again while you hide a different number. Repeat the process, giving different students an opportunity to share their thinking.
9. Repeat the process one more time, but this time use two or three sticky notes to hide two or three different numbers at the same time. (They can be consecutive numbers or numbers that are not consecutive.)
10. Count to three again and ask students to whisper the names of the numbers that are hiding. Then ask, "How do you know?" and call on several students to share their reasoning. Before moving on, place the sticky notes so they are below the numbers they covered (so that the numbers are visible). The sticky notes serve as a reminder of the numbers that were hidden.
11. Now ask, "How can we tell which number is less (the least) and which number is more (the most)?" Use the think-pair-share strategy to bolster students' willingness to share and give everyone the opportunity to share. To do this, tell students to think alone first. Then direct them to share their thinking with a shoulder partner. Provide sentence frames to support their conversations. Next, encourage someone to share with the whole class. Promote active listening by asking another student to repeat what they heard the first student share. Then ask, "Who has a different way to tell which number is less (the least)?" Encourage students to share their different strategies.

Math Talk

Think-Pair-Share

When teachers ask a question to the whole class, often one student responds. However, when teachers use the think-pair-share strategy, all students get a chance to talk about the question. This can be a chance for students to rehearse with a partner what they may want to say to the whole class, or an opportunity for them to predict what will happen in a safe setting. With experience, students become comfortable with the think-pair-share strategy and appreciate the chance to express their ideas before voicing them to the whole class.

Discussion Buddies

Often, students don't know who to turn to when teachers ask them to talk to a partner. To save time and alleviate any issues that might arise, instruct students to share their thinking specifically with a *shoulder partner*.

Teaching Tip

Sentence Frames

Provide the following sentence frames to support student conversation:

I think ____ is less because ________________.
I think ____ is the least because ____________.
I think ____ is more because ______________.
I think ____ is the most because ___________.

12. Now ask, "Who has a different way to tell which number is more (the most)?" Encourage students to share their different strategies.

13. Continue, "You have convinced me that ___ is more than ___." Then ask, "I wonder, how can we use the number path to help us figure out how far apart these two numbers are?"

14. Use the think-pair-share strategy again, providing time for students to think alone and then share with a shoulder partner. As you call on volunteers to share with the whole class, encourage them to come up to the number path and point as they explain their thinking.

15. If no one suggests starting on the larger number and counting numbers to the left to get to the smaller number, encourage this thinking by asking, "Can we find out how far apart the two numbers are if we start on the larger number?" Ask a volunteer to come up to the number path and try.

16. Now cut the number path into thirds and display each third, one under the other. Ask students to watch as you do this. Your number path should look something like this:

1	2	3	4	5	6	7	8	9	10
11	12	13	14	15	16	17	18	19	20
21	22	23	24	25	26	27	28	29	30

Number path labeled *1* through *30* that has been cut into thirds, with the second and third parts placed sequentially below each other.

17. Ask, "What do you notice about the number paths now?" Possible responses may be that the second number in the numbers below is the same as the numbers above. Let several students come up to the display and point to the patterns they see in the arrangement.

A Child's Mind . . .

Talking about *Less/Least* and *More/Most*

The academic language of *more* and *most* may be confusing to young children or second language learners. When comparing two numbers, *more* is used; but, when thinking about the relationship among three or more numbers, *most* is used. This kind of comparison is also true for *less* and *least*; *less* is used when two numbers are compared and *least* is used when three or more numbers are compared. It is important to find ways to make this language explicit with young learners so they become familiar with the different contexts.

A Child's Mind . . .

Counting On

"How far apart are these two numbers?" is a question about the distance between two numbers, which is a measurement idea. Some children with experience using game boards to count from one space to the next will see how they can respond to this question by *counting on* from the smaller number to the larger number. However, often children make the mistake of counting the number they are on as they count to find the difference between two numbers. It's important to model for them that what is being counted is the *space* between the numbers, or the number of jumps it takes to get from one number to the next. It may be helpful to act out the counting by pretending you are taking finger steps on the number path to get from one number to the next. Students are exploring ideas here about subtraction, or finding the difference, by counting up to the total.

18. Ask, "What else do you notice about this arrangement?" Possible responses may be that each number in the middle row has a one in it except for the last number and each number in the bottom row has a two in it except for the last number.

19. Say, "Let's read each of these numbers in order as I point." Count out from one to thirty, pointing to each number.

20. Close the discussion by making the number path and some sticky notes available to students to explore and investigate on their own during independent class times. (This activity could be available as a learning station during math workshop.) Students are likely to use more than two or three sticky notes to challenge each other about the hidden numbers.

Teaching Tip

Displaying a Hundreds Chart

If you have a pocket hundreds chart, this is an appropriate time to display this chart with numbers to thirty. Some of the numbers could be turned over so students could play a guessing game independently or in pairs.

Using a Number Path with Numerals 1–50

Introducing

7

Related Lesson

- L-4 Jumping by Ones and Tens

Key Questions

- How did you know which number is (numbers are) hiding?
- What is another way to know that [number] is behind the sticky note?
- How can we tell which number is less (the least) and which number is more (the most)?

Suggested Grade Level

Grade 1

Materials

number path labeled *1* through *50*, with the numbers spaced equally apart (approximately 3 inches), created using paper sentence strips or cash register paper that is approximately 3 inches tall

large sticky notes, approximately 3-inch-by-3-inch squares

Teaching Directions

1. Prepare a number path per the materials list. Make sure the number path is large enough for everyone in class to see it.

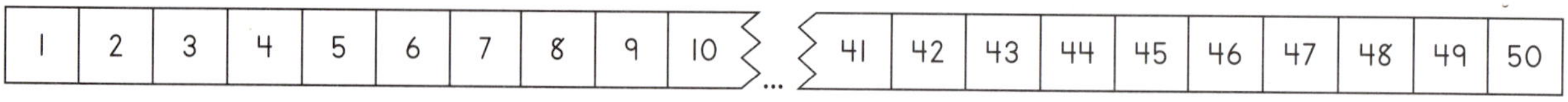

Number path labeled *1* through *50* (abbreviated here in order to fit on this page).

2. Gather students on the rug so that everyone has a good view of the number path. Seat them close together to make listening to each other easier.
3. Draw students' attention to the number path. Say, "This is a number path. I am going to hide one of the numbers on the number path. I want you to figure out which number is hiding."
4. Ask students to look at the floor and cover their eyes. While they are doing this, cover one of the numbers using a sticky note.
5. Instruct students to look up and uncover their eyes. Ask, "Which number is hiding?" Remind students to keep their answer in their head while you give everyone time to think. Ask them to give you a quiet signal, such as a thumbs-up on their chest, when they have an answer to share.
6. Now say, "I am going to count to three. When I say 'three,' I want everyone to whisper the name of the number that is hiding."
7. Call on a student to explain his or her thinking. Ask, "How did you know which number is hiding?" Most students will likely determine the hidden number by noticing what it comes after. For example, "Fourteen is hiding because the sticky note is after thirteen and I know fourteen comes after thirteen." Encourage students to consider

A Child's Mind . . .

Providing Time to Think

Asking students to "whisper" their answer is a helpful way to give all students an opportunity to "say the answer." It also de-emphasizes the significance of saying the answer and allows students to focus more on reasoning and strategies.

which number follows the hidden number by asking, "What is another way to know that fourteen is behind the sticky note?" Remove the sticky note.

8. Ask students to cover their eyes again while you hide a different number. Repeat the process, giving different students an opportunity to share their thinking.
9. Repeat the process one more time, but this time use two or three sticky notes to hide two or three different numbers at the same time. (They can be consecutive numbers or numbers that are not consecutive.)
10. Count to three again and ask students to whisper the names of the two numbers that are hiding. Then ask, "How do you know?" and call on several students to share their reasoning. Before moving on, place the sticky notes so they are below the numbers they covered (so the numbers are visible). The sticky notes serve as a reminder of the numbers that were hidden.
11. Now ask, "How can we tell which number is less (the least) and which number is more (the most)?" Use the think-pair-share strategy to bolster students' willingness to share and give everyone the opportunity to share. To do this, tell students to think alone first. Then direct them to share their thinking with a shoulder partner. Provide sentence frames to support their conversations. Next, encourage someone to share with the whole class. Promote active listening by asking another student to repeat what they heard the first student share. Then ask, "Who has a different way to tell which number is less (the least)?" Encourage students to share their different strategies.
12. Now ask, "Who has a different way to tell which number is more (the most)?" Encourage students to share their different strategies.

Math Talk

Think-Pair-Share

When teachers ask a question to the whole class, often one student responds. However, when teachers use the think-pair-share strategy, all students get a chance to talk about the question. This can be an opportunity for students to rehearse with a partner what they may want to say to the whole class, or a chance for them to predict what will happen in a safe setting. With experience, students become comfortable with the think-pair-share strategy and appreciate the opportunity to express their ideas before voicing them to the whole class.

Discussion Buddies

Often, students don't know who to turn to when teachers ask them to talk to a partner. To save time and alleviate any issues that might arise, instruct students to share their thinking specifically with a *shoulder partner*.

Teaching Tip

Sentence Frames

Provide the following sentence frames to support student conversation:
I think ____ is less because ______________.
I think ____ is the least because ______________.
I think ____ is more because ______________.
I think ____ is the most because ______________.

A Child's Mind . . .

Talking about *Less/Least* and *More/Most*
The academic language of *more* and *most* may be confusing to young children or second language learners. When comparing two numbers, *more* is used; but, when thinking about the relationship among three or more numbers, *most* is used. This kind of comparison is also true for *less* and *least*; *less* is used when two numbers are compared and *least* is used when three or more numbers are compared. It is important to find ways to make this language explicit with young learners so they are familiar with the different contexts.

A Child's Mind . . .

Counting On
"How far apart are these two numbers?" is a question about the distance between two numbers, which is a measurement idea. Some children with experience using game boards to count from one space to the next will see how they can respond to this question by *counting on* from the smaller number to the larger number. However, often children make the mistake of counting the number they are on as they count to find the difference between two numbers. It's important to model for them that what is being counted is the *space* between the numbers, or the number of jumps it takes to get from one number to the next. It may be helpful to act out the counting by pretending you are taking finger steps on the number path to get from one number to the next. Students are exploring ideas here about subtraction, or finding the difference, by counting up to the total.

13. Continue, "You have convinced me that ___ is more than ___." Then ask, "I wonder, how can we use the number path to help us figure out how far apart these two numbers are?"
14. Use the think-pair-share strategy again, providing time for students to think alone and then share with a shoulder partner. As you call on volunteers to share with the whole class, encourage them to come up to the number path and point as they explain their thinking.
15. If no one suggests starting on the larger number and counting numbers to the left to get to the smaller number, encourage this thinking by asking, "Can we find out how far apart the two numbers are if we start on the larger number?" Ask a volunteer to come up to the number path and try.
16. Close the discussion by making the number path and some sticky notes available to students to explore and investigate on their own during independent class times. (This activity could be available as a learning station in a math workshop.) Students are likely to use more than two or three sticky notes to challenge each other about the hidden numbers.

Teaching Tip

Displaying a Hundreds Chart
If you have a pocket hundreds chart, this is an appropriate time to display this chart with numbers to fifty. Some of the numbers could be turned over so students could play a guessing game independently or in pairs.

Using a Number Path with Numerals 31–60

Introducing

8

Related Game

- G-2 Mystery Number (Version 2)

Key Questions

- How did you know which number is (numbers are) hiding?
- What is another way to know that [number] is behind the sticky note?
- How can we tell which number is less (the least) and which number is more (the most)?

Suggested Grade Level

Grade 1

Materials

number path labeled *31* through *60*, with the numbers spaced equally apart (approximately 3 inches), created using paper sentence strips or cash register paper that is approximately 3 inches tall

large sticky notes, approximately 3-inch-by-3-inch squares

Teaching Directions

1. Prepare a number path per the materials list. Make sure the number path is large enough for everyone in class to see it.

31	32	33	34	35	36	37	38	39	40	…	51	52	53	54	55	56	57	58	59	60

Number path labeled *31* through *60* (abbreviated here in order to fit on this page).

2. Gather students on the rug so that everyone has a good view of the number path. Seat them close together to make listening to each other easier.

3. Draw students' attention to the number path. Say, "This is part of a number path. What parts are missing?" Most likely students will notice that it starts at thirty-one instead of one and continues to sixty.

4. Now say, "I am going to hide one of the numbers on the number path. I want you to figure out which number is hiding."

5. Ask students to look at the floor and cover their eyes. While they are doing this, cover one of the numbers with a sticky note.

6. Instruct students to look up and uncover their eyes. Ask, "Which number is hiding?" Remind students to keep their answer in their head while you give everyone time to think. Ask them to give you a quiet signal, such as a thumbs-up on their chest, when they have an answer to share.

7. Now say, "I am going to count to three. When I say 'three,' I want everyone to whisper the name of the number that is hiding."

8. Call on a student to explain their thinking. Ask, "How did you know which number is hiding?" Most students will likely determine the hidden number by noticing what

A Child's Mind . . .

Providing Time to Think
Asking students to "whisper" their answer is a helpful way to give all students an opportunity to "say the answer." It also de-emphasizes the significance of saying the answer and allows students to focus more on reasoning and strategies.

it comes after. For example, "Fifty-three is hiding because the sticky note is after fifty-two, and I know fifty-three comes after fifty-two." Encourage students to consider which number follows the hidden number by asking, "What is another way to know that fifty-three is behind the sticky note?" Possible responses: "It is just before fifty-four" or "When you count, you say fifty-one, fifty-two, fifty-three, fifty-four, fifty-five." Remove the sticky note.

9. Ask students to cover their eyes again while you hide a different number. Repeat the process, giving different students an opportunity to share their thinking.
10. Repeat the process one more time, but this time use two or three sticky notes to hide two or three different numbers at the same time. (They can be consecutive numbers or numbers that are not consecutive.)
11. Count to three again and ask students to whisper the names of the numbers that are hiding. Then ask, "How do you know?" and call on several students to share their reasoning. Before moving on, place the sticky notes so they are below the numbers they covered (so the numbers are visible). The sticky notes serve as a reminder of the numbers that were hidden.
12. Now ask, "How can we tell which number is less (the least) and which number is more (the most)?" Use the think-pair-share strategy to bolster students' willingness to share and give everyone the opportunity to share. To do this, tell students to think alone first. Then direct them to share their thinking with a shoulder partner. Provide sentence frames to support their conversations. Next, encourage someone to share with the whole class. Promote active listening by asking another student to repeat what they heard the first student share. Then ask, "Who has

Math Talk

Think-Pair-Share

When teachers ask a question to the whole class, often one student responds. However, when teachers use the think-pair-share strategy, all students get a chance to talk about the question. This can be an opportunity for students to rehearse with a partner what they may want to say to the whole class, or a chance for them to predict what will happen in a safe setting. With experience, students become comfortable with the think-pair-share strategy and appreciate the opportunity to express their ideas before voicing them to the whole class.

Discussion Buddies

Often, students don't know who to turn to when teachers ask them to talk to a partner. To save time and alleviate any issues that might arise, instruct students to share their thinking specifically with a *shoulder partner*.

Teaching Tip

Sentence Frames

Provide the following sentence frames to support student conversation:

I think ____ is less because ______________.
I think ____ is the least because ____________.
I think ____ is more because ______________.
I think ____ is the most because ___________.

A Child's Mind . . .

Talking about *Less/Least* and *More/Most*

The academic language of *more* and *most* may be confusing to young children or second language learners. When comparing two numbers, *more* is used; but, when thinking about the relationship among three or more numbers, *most* is used. This kind of comparison is also true for *less* and *least*; *less* is used when two numbers are compared and *least* is used when three or more numbers are compared. It is important to find ways to make this language explicit with young learners so they are familiar with the different contexts.

A Child's Mind . . .

Counting On

"How far apart are these two numbers?" is a question about the distance between two numbers, which is a measurement idea. Some children with experience using game boards to count from one space to the next will see how they can respond to this question by *counting on* from the smaller number to the larger number. However, often children make the mistake of counting the number they are on as they count to find the difference between two numbers. It's important to model for them that what is being counted is the *space* between the numbers, or the number of jumps it takes to get from one number to the next. It may be helpful to act out the counting by pretending you are taking finger steps on the number path to get from one number to the next. Students are exploring ideas here about subtraction, or finding the difference, by counting up to the total.

a different way to tell which number is less (least)?" Encourage students to share their different strategies.

13. Now ask, "Who has a different way to tell which number is more (the most)?" Encourage students to share their different strategies.
14. Continue, "You have convinced me that ___ is more than ___." Then ask, "I wonder, how can we use the number path to help us figure out how far apart these two numbers are?"
15. Use the think-pair-share strategy again, providing time for students to think alone and then share with a shoulder partner. As you call on volunteers to share with the whole class, encourage them to come up to the number path and point as they explain their thinking.
16. If no one suggests starting on the larger number and counting numbers to the left to get to the smaller number, encourage this thinking by asking, "Can we find out how far apart the two numbers are if we start on the larger number?" Ask a volunteer to come up to the number path and try.
17. Close the discussion by making the number path and some sticky notes available to students to explore and investigate on their own during independent class times. (This activity could be available as a learning station in math workshop.) Students are likely to use more than two or three sticky notes to challenge each other about the hidden numbers.

Teaching Tip

Displaying a Hundreds Chart

If you have a pocket hundreds chart, this is an appropriate time to display this chart with numbers thirty-one to sixty. Some of the numbers could be turned over so students could play a guessing game independently or in pairs.

Using a Number Path with Numerals 51–100

Introducing

9

Related Game

- G-2 Mystery Number (Version 2)

Key Questions

- How did you know which number is (numbers are) hiding?
- What is another way to know that [number] is behind the sticky note?
- How can we tell which number is less (the least) and which number is more (the most)?

Suggested Grade Level

Grade 1

Materials

number path labeled *51* through *100*, with the numbers spaced equally apart (approximately 3 inches), created using paper sentence strips or cash register paper that is approximately 3 inches tall

large sticky notes, approximately 3-inch-by-3-inch squares

Teaching Directions

1. Prepare a number path per the materials list. Make sure the number path is large enough for everyone in class to see it.

51	52	53	54	55	56	57	58	59	60	…	91	92	93	94	95	96	97	98	99	100

Number path labeled *51* through *100* (abbreviated here in order to fit on this page).

2. Gather students on the rug so that everyone has a good view of the number path. Seat them close together to make listening to each other easier.
3. Draw students' attention to the number path. Say, "This is part of a number path. What parts are missing?" Most likely students will notice that it starts at fifty-one instead of one and continues to one hundred.
4. Now say, "I am going to hide one of the numbers on the number path. I want you to figure out which number is hiding."
5. Ask students to look at the floor and cover their eyes. While they are doing this, cover one of the numbers with a sticky note.
6. Instruct students to look up and uncover their eyes. Ask, "Which number is hiding?" Remind students to keep their answer in their head while you give everyone time to think. Ask them to give you a quiet signal, such as a thumbs-up on their chest, when they have an answer to share.
7. Now say, "I am going to count to three. When I say 'three,' I want everyone to whisper the name of the number that is hiding."
8. Call on a student to explain their thinking. Ask, "How did you know which number is hiding?" Most students will likely determine the hidden number by

A Child's Mind . . .

Providing Time to Think

Asking students to "whisper" their answer is a helpful way to give all students an opportunity to "say the answer." It also de-emphasizes the significance of saying the answer and allows students to focus more on reasoning and strategies.

noticing what it comes after. For example, "Fifty-three is hiding because the sticky note is after fifty-two and I know fifty-three comes after fifty-two." Encourage students to consider which number follows the hidden number by asking, "What is another way to know that fifty-three is behind the sticky note?" Possible responses: "It's just before fifty-four" or "When you count, you say fifty-one, fifty-two, fifty-three, fifty-four, fifty-five." Remove the sticky note.

9. Ask students to cover their eyes again while you hide a different number. Repeat the process, giving different students an opportunity to share their thinking.

10. Repeat the process one more time, but this time use two or three sticky notes to hide two or three different numbers at the same time. (They can be consecutive numbers or numbers that are not consecutive.)

11. Count to three again and ask students to whisper the names of the numbers that are hiding. Then ask, "How do you know?" and call on several students to share their reasoning. Before moving on, place the sticky notes so they are below the numbers they covered (so the numbers are visible). The sticky notes serve as a reminder of the numbers that were hidden.

12. Now ask, "How can we tell which number is less (the least) and which number is more (the most)?" Use the think-pair-share strategy to bolster students' willingness to share and give everyone the opportunity to share. To do this, tell students to think alone first. Then direct them to share their thinking with a shoulder partner. Provide sentence frames to support their conversations. Next, encourage someone to share with the whole class. Promote active listening by asking another student to repeat what they heard the first student share. Then ask, "Who has

Math Talk

Think-Pair-Share

When teachers ask a question to the whole class, often one student responds. However, when teachers use the think-pair-share strategy, all students get a chance to talk about the question. This can be a chance for students to rehearse with a partner what they may want to say to the whole class, or an opportunity for them to predict what will happen in a safe setting. With experience, students become comfortable with the think-pair-share strategy and appreciate the chance to express their ideas before voicing them to the whole class.

Discussion Buddies

Often, students don't know who to turn to when teachers ask them to talk to a partner. To save time and alleviate any issues that might arise, instruct students to share their thinking specifically with a *shoulder partner*.

Teaching Tip

Sentence Frames

Provide the following sentence frames to support student conversation:

I think ____ is less because ______________.
I think ____ is the least because ___________.
I think ____ is more because _____________.
I think ____ is the most because ___________.

a different way to tell which number is less (the least)?" Encourage students to share their different strategies.

13. Now ask, "Who has a different way to tell which number is more (the most)?" Encourage students to share their different strategies.
14. Continue, "You have convinced me that ___ is more than ___." Then ask, "I wonder, how can we use the number path to help us figure out how far apart these two numbers are?"
15. Use the think-pair-share strategy again, providing time for students to think alone and then share with a shoulder partner. As you call on volunteers to share with the whole class, encourage them to come up to the number path and point as they explain their thinking.
16. If no one suggests starting on the larger number and counting numbers to the left to get to the smaller number, encourage this thinking by asking, "Can we find out how far apart the two numbers are if we start on the larger number?" Ask a volunteer to come up to the number path and try.
17. Close the discussion by making the number path and some sticky notes available to students to explore and investigate on their own during independent class times. (This activity could be available as a learning station in math workshop.) Students are likely to use more than two or three sticky notes to challenge each other about the hidden numbers.

A Child's Mind . . .

Talking about *Less/Least* and *More/Most*

The academic language of *more* and *most* may be confusing to young children or second language learners. When comparing two numbers, *more* is used; but, when thinking about the relationship among three or more numbers, *most* is used. This kind of comparison is also true for *less* and *least*; *less* is used when two numbers are compared and *least* is used when three or more numbers are compared. It is important to find ways to make this language explicit with young learners so they are familiar with the different contexts.

A Child's Mind . . .

Counting On

"How far apart are these two numbers?" is a question about the distance between two numbers, which is a measurement idea. Some children with experience using game boards to count from one space to the next will see how they can respond to this question by *counting on* from the smaller number to the larger number. However, often children make the mistake of counting the number they are on as they count to find the difference between two numbers. It's important to model for them that what is being counted is the *space* between the numbers, or the number of jumps it takes to get from one number to the next. It may be helpful to act out the counting by pretending you are taking finger steps on the number path to get from one number to the next. Students are exploring ideas here about subtraction, or finding the difference, by counting up to the total.

Teaching Tip

Displaying a Hundreds Chart

If you have a pocket hundreds chart, this is an appropriate time to display this chart with numbers fifty-one to one hundred. Some of the numbers could be turned over so students could play a guessing game independently or in pairs.

Using an Open Number Line with Numerals 0 and 50

Introducing

10

Related Game

- G-2 Mystery Number (Version 2)

Key Questions

- Why do you think this is called an open number line?
- Why do you think [number] is there on the number line?
- How can we find out how far it is from [number] to [number] on a number line?
- What are the different ways you can make jumps on a number line? Which way do you like best? Why?

Suggested Grade Level

Grade 1

Materials

open number line labeled *0* and *50*

Teaching Directions

1. Prepare an open number line per the materials list. Make sure the number line is large enough for everyone in class to see it.

Open number line with numbers *0* and *50*.

2. Gather students on the rug so that everyone has a good view of the number line. Seat them close together to make listening to each other easier.
3. Draw students' attention to the number line. Remind students that it is called an *open* (or *empty*) *number line*. Ask, "Why do you think this is called an open number line?" Most likely, students will respond with the idea that some of the numbers are missing or that only two numbers are shown.
4. Now ask, "What do the little arrows by zero and fifty mean?" The response might be that the numbers can get smaller going to the left and keep on going beyond fifty to the right.
5. Ask, "Who can point to where you think twenty-five would be on this line?" Have a student volunteer indicate where they think twenty-five would be on the line. Ask, "Why do you think this?" Before marking *25*, check with the rest of the students to see if they agree with the location for twenty-five. If there are other suggestions, have students come up and point to the location and

Technology Tip

Using an Interactive Whiteboard
An interactive whiteboard can also be used to display an open number line. Use the line option to draw a straight line, then use the pen to label each end.

Teaching Tip

Reviewing Vocabulary
Rather than assume that children know the meaning of words used in math, take a minute or two to be sure that everyone has a common understanding of the vocabulary you'll be using during a lesson.

Teaching Tip

Making Connections
Some children may not make sense of the idea that there are numbers that can be less than zero. If this comes up, refer to the thermometer that sometimes dips below zero when the weather is very cold.

explain why they think twenty-five belongs there. Students will most likely note that twenty-five is halfway between zero and fifty.

6. Ask, "Who can point to where you think three would be on this line?" Have a student volunteer indicate where they think three would be on the line. Ask, "Why do you think this?" Before marking *3*, check with the rest of the students to see if they agree with the location for three. If there are other suggestions, have students come up and point to the location and explain why they think three belongs there. Students will probably say that three is close to zero.

7. Now ask, "Who can point to where you think forty-six would be on this line?" Have a student volunteer indicate where he or she thinks forty-six would be on the line. Ask, "Why do you think this?" Before marking *46*, check with the rest of the students to see if they agree with the location for forty-six. If there are other suggestions, have students come up and point to the location and explain why they think forty-six belongs there. Students will most likely note that forty-six is pretty close to fifty.

8. Gesturing from the 3 to 25, ask, "How far apart are these two numbers?" Say, "Let's use what we know about friendly numbers such as five and ten to help us."

9. Ask, "How far would it be to jump from three to five?" An example of student thinking could be to jump by two. Record the jump of *+2* as shown.

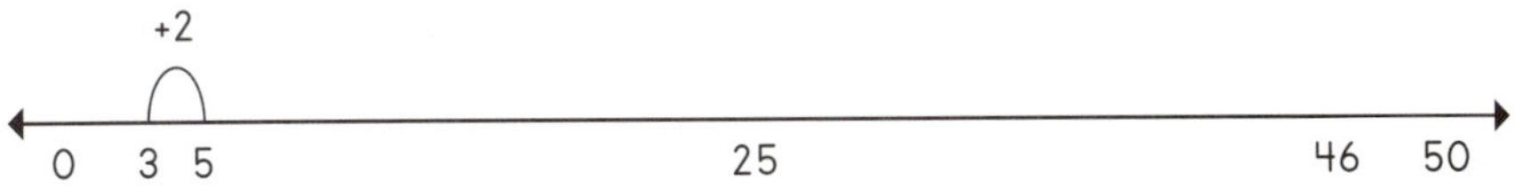

Recording 1: "How far would it be to jump from three to five?"

Teaching Tip

Being Comfortable with Approximations

Do not give students the option of using measuring tools when locating numbers on the number line during this activity. In this instance, we want students to begin to feel comfortable with approximation, using what they know about the relative size of the numbers. Instead of counting squares, as you would on a number path, you want students to be able to move by imaginary jumps of squares along the number line.

10. Ask, "Is there a way to know how far it is from five to twenty-five?" Possible responses might be to count by fives or tens. Some students may just know that twenty-five is twenty more than five. Agree and let students know that sometimes we can make large jumps or successive smaller jumps to show the distance. Both ways work. Record *+20* or jumps of *+5*, depending on which method your students suggest.

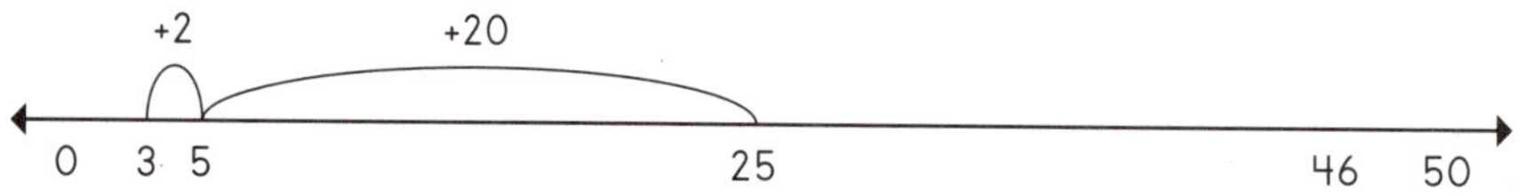

Recording 2: "How far would it be to jump from five to twenty-five?"

Math Matters!

Counting by Ones

Don't be alarmed if a student starts with five and makes jumps of one to get to twenty-five. Many students go back to what they know when they are faced with something they perceive as "hard." Allow them to make jumps of one, then encourage them to use their knowledge of fives and tens to jump to twenty-five. This teaching strategy helps students see they can trust using fives and tens without needing to count by ones each time. They will also use more efficient methods after they become more comfortable with the number line.

11. Close with a discussion about the different ways you can make jumps on a number line. Say, "When we are trying to find the difference between two numbers, we can use different ways to jump. Sometimes it makes sense to jump by tens or fives. We can always count by ones, but that might take more time and we might get mixed up when we try to count the jumps."

12. Ask students to tell their shoulder partner the way to jump that they liked best.

Using an Open Number Line with Numerals 0 and 100

Introducing

11

Related Game

- G-2 Mystery Number (Version 2)

Key Questions

- Why do you think this is called an open number line?
- Why do you think [number] is there on the number line?
- How can we find out how far it is from [number] to [number] on the number line?
- What are the different ways you can make jumps on a number line? What way do you like best? Why?

Suggested Grade Level

Grade 1

Materials

open number line labeled *0* and *100*

Teaching Directions

1. Prepare an open number line per the materials list. Make sure the number line is large enough for everyone in class to see it.

Open number line with numbers *0* and *100*.

2. Gather students on the rug so that everyone has a good view of the number line. Seat them close together to make listening to each other easier.
3. Draw students' attention to the number line. Remind students that it is called an *open* (or *empty*) *number line.* Ask, "Why do you think this is called an open number line?" Most likely, students will respond with the idea that some of the numbers are missing or that only two numbers are shown.
4. Now ask, "What do the little arrows by zero and one hundred mean?" The response might be that the numbers get smaller going to the left and keep on going beyond one hundred to the right.
5. Ask, "Who can point to where you think fifty would be on this line?" Have a student volunteer indicate where they think fifty would be on the line. Ask, "Why do you think this?" Before marking *50*, check with the rest of the students to see if they agree with the location for fifty. If there are other suggestions, have students come up and point to the location and explain why they think fifty belongs there. Students will most likely note that fifty is halfway between zero and one hundred.

Technology Tip

Using an Interactive Whiteboard

An interactive whiteboard can also be used to display an open number line. Use the line option to draw a straight line, then use the pen to label each end.

Teaching Tip

Reviewing Vocabulary

Rather than assume that children know the meaning of words used in math, take a minute or two to be sure that everyone has a common understanding of the vocabulary you'll be using during a lesson.

Teaching Tip

Making Connections

Some children may not make sense of the idea that there are numbers that can be less than zero. If this comes up, refer to the thermometer that sometimes dips below zero when the weather is very cold.

6. Ask, "Who can point to where you think ten would be on this line?" Have a student volunteer indicate where they think ten would be on the line. Ask, "Why do you think this?" Before marking *10*, check with the rest of the students to see if they agree with the location for ten. If there are other suggestions, have students come up and point to the location and tell why they think ten belongs there. Students might note that half of twenty-five is about twelve, so ten would be less than twelve. It is less likely that someone will try to divide up the section into five approximately equal parts.

7. Now ask, "Who can point to where you think ninety-six would be on this line?" Have a student volunteer indicate where they think ninety-six would be on the line. Ask, "Why do you think this?" Before marking *96*, check with the rest of the students to see if they agree with the location for ninety-six. If there are other suggestions, have students come up and point to the location and explain why they think ninety-six belongs there. Students will most likely note that ninety-six is pretty close to one hundred.

8. Gesturing from 10 to 50, ask, "How far apart are these two numbers?" Say, "I think we can use what we know about groups of ten to help us answer this question."

9. Ask, "How far would it be to jump from ten to fifty?" An example of student thinking could be by making one jump of forty. Record the jump of *+40* as shown.

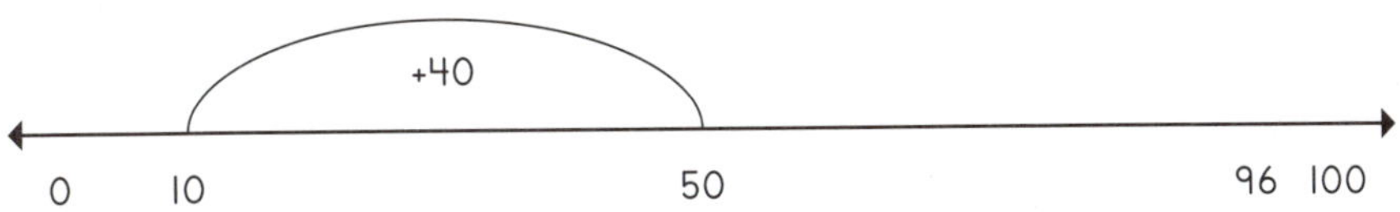

Recording 1: "How far would it be to jump from ten to fifty?"

Teaching Tip

Being Comfortable with Approximations

Do not give students the option of using measuring tools when locating numbers on the number line during this activity. In this instance, we want students to begin to feel comfortable with approximation, using what they know about the relative size of the numbers. Instead of counting squares, like on a number path, we want students to be able to move by imaginary jumps of squares along the number line.

10. Ask, "Is there another way to know how far it is from ten to fifty?" A possible response might be to count by tens. Agree and let students know that sometimes we can make large jumps or successive smaller jumps to show the distance. Both ways work.

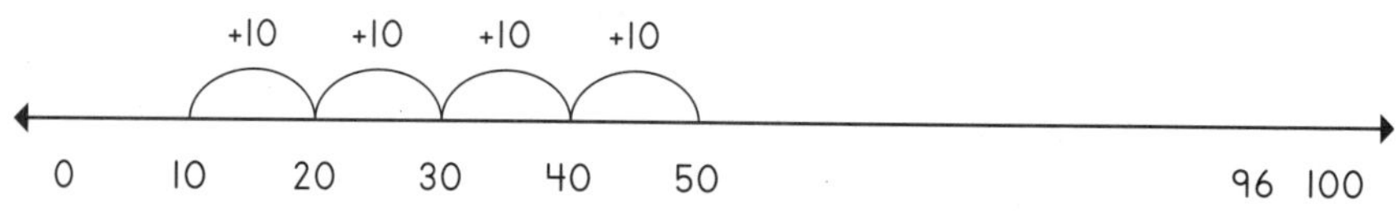

Recording 2: "How far would it be to jump from ten to fifty?"

11. Ask, "If we wanted to now jump to from fifty to ninety-six, how far would that be?" Some students will know it is a jump of forty-six. Others will jump from fifty to sixty to seventy to eighty to ninety and then jump six more steps. Record what the students say. If there are two ways, record both of them. Reaffirm, "So how far did we jump from fifty to ninety-six?" The expected response is forty-six.

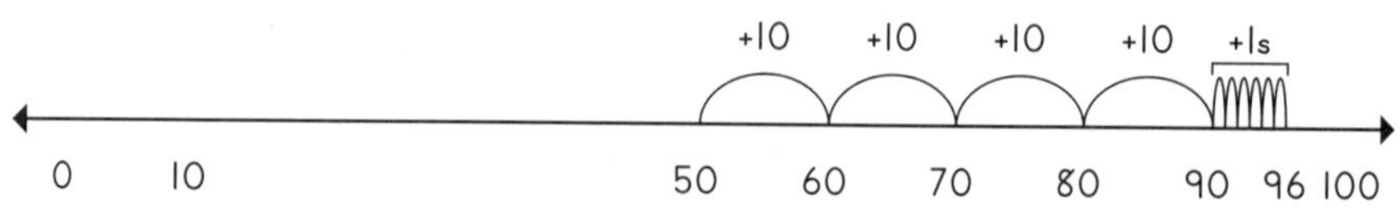

Recording 3: "How far would it be to jump from fifty to ninety-six?"

12. Close with a discussion about the different ways you can make jumps on a number line. Say, "When we are trying to find the difference between two numbers, we can use different ways to jump. Sometimes

it makes sense to jump by tens or fives. We can always count by ones, but that might take more time and we might get mixed up when we try to count the jumps."

13. Ask students to tell their shoulder partner the way to jump that they liked best.

Math Matters!

Counting by Ones

Don't be alarmed if a student starts with ten and then makes jumps of one. Many students go back to what they know when they are faced with something they perceive as "hard." Allow them to make jumps of one, then encourage them to use their knowledge of fives and tens to jump to fifty. This teaching strategy helps them see they can trust using fives and tens without needing to count by ones each time. Students will also use more efficient methods after they become more comfortable with the number line.

Introducing 12

Using an Open Number Path without Numerals 1–60

Suggested Grade Level

Grade 2

Materials

open number path, created using red and yellow paper squares, 30 of each color, organized in groups of five starting with red. (See Reproducible A, How to Make an Open Number Path.)

Related Lesson

- L-3 Building the Open Number Path

Key Questions

- How did you know there were sixty?
- Do you think there are the same number of red squares as yellow squares? How can you find out?
- How can you tell how many red and yellow squares there are without counting them one by one?
- How do you know that red squares come next?

Teaching Directions

1. Prepare an open number path per the materials list. Make sure the number path is large enough for everyone in class to see it.

Open number path (abbreviated here in order to fit on this page).

2. Gather students on the rug so that everyone has a good view of the number path. Seat them close together to make listening to each other easier.
3. Draw students' attention to the number path. Say, "This is a number path. I made this number path with some red and yellow squares." Ask students to look at the path and think about the question: How many red and yellow squares are there?
4. Remind students to keep their answer in their head while everyone gets time to think. Ask them to give you a quiet signal, such as a thumbs-up on their chest, when they have an answer to share.
5. Now share with students, "I am going to count to three. When I say 'three,' I want everyone to whisper the number of squares that you see." Students will probably respond with sixty, counting groups of fives or tens.
6. Call on a student to explain their thinking. Ask, "How did you know there were sixty?" It is likely that students will come up and count each square, tapping the squares with their fingers.
7. If students do not seem to be able to offer a strategy other than counting each square, suggest looking at the groups of five red and five yellow squares. Say, "Let's count

A Child's Mind . . .

Providing Time to Think

Asking students to "whisper" their answer is a helpful way to give all students an opportunity to "say the answer." It also de-emphasizes the significance of saying the answer and allows students to focus more on reasoning and strategies.

A Child's Mind . . .

Touching the Number Path

When reading, children start out by touching each word as they read it. Later, they begin to move their finger under a line of text as they read more fluently. So, too, in math do children begin by touching each number, one at a time, as they count on a number path or number line. As their thinking develops, they become more comfortable counting in their head.

by fives." Place your hands like parentheses around each group of five red squares and five yellow squares as you count.

8. Continue, "Now I'd like you to think about how many squares there are in half of this path." Remind students again to keep their answer in their head while you give everyone time to think. Ask them to give you a quiet signal, such as a thumbs-up on their chest, when they have an answer to share.
9. Count again to three and repeat the process, giving different students an opportunity to share their thinking. Encourage students to think about how the path was made. Mimic counting out five red squares, gluing them down, then counting out five yellow squares and gluing them down, then five more red squares, and then five more yellow squares, and so forth.
10. It is helpful if students notice there are the same number of red squares as yellow squares. If this does not come up, ask, "Do you think there are the same number of red squares as yellow squares? How can you find out?" Allow a couple of students to share their thinking.
11. Now fold the number path in half so that students can only see thirty squares (fifteen red and fifteen yellow). Have students watch as you do this.
12. Point to the folded number path and say, "Think to yourself: How many red and yellow squares are in this half? Give me a quiet signal when you have your answer."
13. Ask students to whisper their answers after you count to three. Students are likely to respond with "Thirty" or "Ten and ten and ten."

14. Ask, "How can you tell how many red and yellow squares there are in this half without counting them one by one?" Use the think-pair-share strategy to bolster students' willingness to share and give everyone the opportunity to share. To do this, tell students to think alone first. Then direct them to share their thinking with a shoulder partner. Next, encourage someone to share with the whole class. Promote active listening by asking another student to repeat what they heard the first student share. Then ask, "Who has a different way to tell how many squares are in this part without counting them one by one?" A possible response would be "Thirty, because thirty and thirty is sixty." Some students might know this because they know three and three is six, so three tens and three tens make six tens.

15. Now unfold the path so students see all sixty squares again. Remind students, "Here we have our sixty squares. Thirty are red and thirty are yellow. What if we wanted to make this path longer and glue on more squares? What color should we use next?"

16. Ask, "How do you know that red squares come next?"

17. Bring the discussion to a close by asking, "How many squares would we then have in all if we added five more red squares and five more yellow squares?" This question challenges students to think abstractly because they are being asked for the total number if ten more were to be added to sixty, but the physical representation has not been altered.

Math Talk

Think-Pair-Share
When teachers ask a question to the whole class, often one student responds. However, when teachers use the think-pair-share strategy, all students get a chance to talk about the question. This can be a chance for students to rehearse with a partner what they may want to say to the whole class, or an opportunity for them to predict what will happen in a safe setting. With experience, students become comfortable with the think-pair-share strategy and appreciate the chance to express their ideas before voicing them to the whole class.

Discussion Buddies
Often, students don't know who to turn to when teachers ask them to talk to a partner. To save time and alleviate any issues that might arise, instruct students to share their thinking specifically with a *shoulder partner*.

Teaching Tip

Keeping the Number Path on Display
It is helpful to keep this path on display in an accessible place, so children can go to it and count the squares again and again. Adding on to the original sixty squares is a project that some students find very engaging. Making a path that is one-hundred squares long is a nice goal for second grade.

Introducing 13 Using a Number Path with Numerals 100–150

Suggested Grade Level

Grade 2

Materials

number path labeled *100* through *150*, with the numbers spaced equally apart (approximately 3 inches), created using paper sentence strips or cash register paper that is approximately 3 inches tall

large sticky notes, approximately 3-inch-by-3-inch squares

Related Game

- G-3 Mystery Number (Version 3)

Key Questions

- How did you know which number is (numbers are) hiding?
- What is another way to know that [number] is behind the sticky note?
- How can we tell which number is less (the least) and which number is more (the most)?

Teaching Directions

1. Prepare a number path per the materials list. Make sure the number path is large enough for everyone in class to see it.

100	101	102	103	104	105	106	107	108	109	…	141	142	143	144	145	146	147	148	149	150

Number path labeled *100* through *150* (abbreviated here in order to fit on this page).

2. Gather students on the rug so that everyone has a good view of the number path. Seat them close together to make listening to each other easier.
3. Draw students' attention to the number path. Say, "This is part of a number path. What parts are missing?" Most likely, students will notice that it starts at 100 instead of 1 and continues to 150.
4. Now say, "I am going to hide one of the numbers on the number path. I want you to figure out which number is hiding."
5. Ask students to look at the floor and cover their eyes. While they are doing this, cover one of the numbers with a sticky note.
6. Instruct students to look up and uncover their eyes. Ask, "Which number is hiding?" Remind students to keep their answer in their head while you give everyone time to think. Ask them to give you a quiet signal, such as a thumbs-up on their chest, when they have an answer to share.
7. Now say, "I am going to count to three. When I say 'three,' I want everyone to whisper the name of the number that is hiding."
8. Call on a student to explain their thinking. Ask, "How did you know which number is hiding?" Most students will likely determine the hidden number by noticing what it comes after. For example, "One hundred

A Child's Mind . . .

Providing Time to Think

Asking students to "whisper" their answer is a helpful way to give all students an opportunity to "say the answer." It also de-emphasizes the significance of saying the answer and allows students to focus more on reasoning and strategies.

twenty-three is hiding because the sticky note is after one hundred twenty-two and I know one hundred twenty-three comes after one hundred twenty-two." Encourage students to consider which number follows the hidden number by asking, "What is another way to know that one hundred twenty-three is behind the sticky note?" Possible responses: "It's just before one hundred twenty-four" or "When you count, you say one hundred twenty-one, one hundred twenty-two, one hundred twenty-three, one hundred twenty-four, one hundred twenty-five." Remove the sticky note.

9. Ask students to cover their eyes again while you hide a different number. Repeat the process, giving different students an opportunity to share their thinking.

10. Repeat the process one more time, but this time use two or three sticky notes to hide two or three different numbers at the same time. (They can be consecutive numbers or numbers that are not consecutive.)

11. Count to three again and ask students to whisper the names of the numbers that are hiding. Then ask, "How do you know?" and call on several students to share their reasoning. Before moving on, place the sticky notes so they are below the numbers they covered (so the numbers are visible). The sticky notes serve as a reminder of the numbers that were hidden.

12. Now ask, "How can we tell which number is less (the least) and which number is more (the most)?" Use the think-pair-share strategy to bolster students' willingness to share and give everyone the opportunity to share. To do this, tell students to think alone first. Then direct them to share their thinking with a shoulder partner. Provide sentence frames to support their conversations. Next,

Math Talk

Think-Pair-Share

When teachers ask a question to the whole class, often one student responds. However, when teachers use the think-pair-share strategy, all students get a chance to talk about the question. This can be a chance for students to rehearse with a partner what they may want to say to the whole class, or an opportunity for them to predict what will happen in a safe setting. With experience, students become comfortable with the think-pair-share strategy and appreciate the chance to express their ideas before voicing them to the whole class.

Discussion Buddies

Often, students don't know who to turn to when teachers ask them to talk to a partner. To save time and alleviate any issues that might arise, instruct students to share their thinking specifically with a *shoulder partner*.

Teaching Tip

Sentence Frames

Provide the following sentence frames to support student conversation:

I think ____ is less because ______________.
I think ____ is the least because ____________.
I think ____ is more because _____________.
I think ____ is the most because ___________.

encourage someone to share with the whole class. Promote active listening by asking another student to repeat what they heard the first student share. Then ask, "Who has a different way to tell which number is less (least)?" Encourage students to share their different strategies.

13. Now ask, "Who has a different way to tell which number is more (the most)?" Encourage students to share their different strategies.

14. Continue, "You have convinced me that ___ is more than ___." Then ask, "I wonder, how can we use the number path to help us figure out how far apart these two numbers are?"

15. Use the think-pair-share strategy again, providing time for students to think alone and then share with a shoulder partner. As you call on volunteers to share with the whole class, encourage them to come up to the number path and point as they explain their thinking.

16. If no one suggests starting on the larger number and counting numbers to the left to get to the smaller number, encourage this thinking by asking, "Can we find out how far apart the two numbers are if we start on the larger number?" Ask a volunteer to come up to the number path and try.

17. Close the discussion by making the number path and some sticky notes available to students to explore and investigate on their own during independent class times. (This activity could be available as a learning station in math workshop.) Students are likely to use more than two or three sticky notes to challenge each other about the hidden numbers.

A Child's Mind . . .

Talking about *Less/Least* and *More/Most*

The academic language of *more* and *most* may be confusing to young children or second language learners. When comparing two numbers, *more* is used; but, when thinking about the relationship among three or more numbers, *most* is used. This kind of comparison is also true for *less* and *least*; *less* is used when two numbers are compared and *least* is used when three or more numbers are compared. It is important to find ways to make this language explicit with young learners so they are familiar with the different contexts.

A Child's Mind . . .

Counting On

"How far apart are these two numbers?" is a question about the distance between two numbers, which is a measurement idea. Some children with experience using game boards to count from one space to the next will see how they can respond to this question by *counting on* from the smaller number to the larger number. However, often children make the mistake of counting the number they are on as they count to find the difference between two numbers. It's important to model for them that what is being counted is the *space* between the numbers, or the number of jumps it takes to get from one number to the next. It may be helpful to act out the counting by pretending you are taking finger steps on the number path to get from one number to the next. Students are exploring ideas here about subtraction, or finding the difference, by counting up to the total.

Introducing 14 Using a Number Path with Numerals 200–250

Suggested Grade Level

Grade 2

Materials

number path labeled *200* through *250*, with the numbers spaced equally apart (approximately 3 inches), created using paper sentence strips or cash register paper that is approximately 3 inches tall

large sticky notes, approximately 3-inch-by-3-inch squares

Related Game

- G-3 Mystery Number (Version 3)

Key Questions

- How did you know which number is (numbers are) hiding?
- What is another way to know that [number] is behind the sticky note?
- How can you tell which number is less (the least) and which number is more (the most)?

Teaching Directions

1. Prepare a number path per the materials list. Make sure the number path is large enough for everyone in class to see it.

200	201	202	203	204	205	206	207	208	209	…	241	242	243	244	245	246	247	248	249	250

Number path labeled *200* through *250* (abbreviated here in order to fit on this page).

2. Gather students on the rug so that everyone has a good view of the number path. Seat them close together to make listening to each other easier.
3. Draw students' attention to the number path. Say, "This is part of a number path. What parts are missing?" Most likely students will notice that it starts at 200 instead of one and continues to 250.
4. Now say, "I am going to hide one of the numbers on the number path. I want you to figure out which number is hiding."
5. Ask students to look at the floor and cover their eyes. While they are doing this, cover one of the numbers with a sticky note.
6. Instruct students to look up and uncover their eyes. Ask, "Which number is hiding?" Remind students to keep their answer in their head while you give everyone time to think. Ask them to give you a quiet signal, such as a thumbs-up on their chest, when they have an answer to share.
7. Now say, "I am going to count to three. When I say 'three,' I want everyone to whisper the name of the number that is hiding."
8. Call on students to explain their thinking. Ask, "How did you know which number is hiding?" Most students will likely determine the hidden number by noticing what it comes after. For example, "Two hundred

A Child's Mind . . .

Numbers Greater Than 200

Second graders may be familiar with the numbers less than one hundred, but they often have not had experiences looking for patterns and relationships with numbers greater than 200.

A Child's Mind . . .

Providing Time to Think

Asking students to "whisper" their answer is a helpful way to give all students an opportunity to "say the answer." It also de-emphasizes the significance of saying the answer and allows students to focus more on reasoning and strategies.

twenty-three is hiding because the sticky note is after two hundred twenty-two and I know two hundred twenty-three comes after two hundred twenty-two." Encourage students to consider which number follows the hidden number by asking, "What is another way to know that two hundred twenty-three is behind the sticky note?" Possible responses: "It's just before two hundred twenty-four" or "When you count, you say two hundred twenty-one, two hundred twenty-two, two hundred twenty-three, two hundred twenty-four, two hundred twenty-five." Remove the sticky note.

9. Ask students to cover their eyes again while you hide a different number. Repeat the process, giving different students an opportunity to share their thinking.

10. Repeat the process one more time, but this time use two or three sticky notes to hide two or three different numbers at the same time. (They can be consecutive numbers or numbers that are not consecutive.)

11. Count to three again and ask students to whisper the names of the numbers that are hiding. Then ask, "How do you know?" and call on several students to share their reasoning. Before moving on, place the sticky notes so they are below the numbers they covered (so the numbers are visible). The sticky notes serve as a reminder of the numbers that were hidden.

12. Now ask, "How can we tell which number is less (the least) and which number is more (the most)?" Use the think-pair-share strategy to bolster students' willingness to share and give everyone the opportunity to share. To do this, tell students to think alone first. Then direct them to share their thinking with a shoulder partner. Provide sentence frames to support their conversations. Next,

Math Talk

Think-Pair-Share

When teachers ask a question to the whole class, often one student responds. However, when teachers use the think-pair-share strategy, all students get a chance to talk about the question. This can be an opportunity for students to rehearse with a partner what they may want to say to the whole class, or a chance for them to predict what will happen in a safe setting. With experience, students become comfortable with the think-pair-share strategy and appreciate the opportunity to express their ideas before voicing them to the whole class.

Discussion Buddies

Often, students don't know who to turn to when teachers ask them to talk to a partner. To save time and alleviate any issues that might arise, instruct students to share their thinking specifically with a *shoulder partner.*

Teaching Tip

Sentence Frames

Provide the following sentence frames to support student conversation:

I think ____ is less because ______________.
I think ____ is the least because ___________.
I think ____ is more because _____________.
I think ____ is the most because __________.

encourage someone to share with the whole class. Promote active listening by asking another student to repeat what they heard the first student share. Then ask, "Who has a different way to tell which number is less (the least)?" Encourage students to share their different strategies.

13. Now ask, "Who has a different way to tell which number is more (the most)?" Encourage students to share their different strategies.
14. Continue, "You have convinced me that ___ is more than ___." Then ask, "I wonder, how can we use the number path to help us figure out how far apart these two numbers are?"
15. Use the think-pair-share strategy again, providing time for students to think alone and then share with a shoulder partner. As you call on volunteers to share with the whole class, encourage them to come up to the number path and point as they explain their thinking.
16. If no one suggests starting on the larger number and counting numbers to the left to get to the smaller number, encourage this thinking by asking, "Can we find out how far apart the two numbers are if we start on the larger number?" Ask a volunteer to come up to the number path and try.
17. Close the discussion by making the number path and some sticky notes available to students to explore and investigate on their own during independent class times. (This activity could be available as a learning station in math workshop.) They are likely to use more than two or three sticky notes to challenge each other about the hidden numbers.

A Child's Mind . . .

Talking about *Less/Least* and *More/Most*

The academic language of *more* and *most* may be confusing to young children or second language learners. When comparing two numbers, *more* is used; but, when thinking about the relationship among three or more numbers, *most* is used. This kind of comparison is also true for *less* and *least*; *less* is used when two numbers are compared and *least* is used when three or more numbers are compared. It is important to find ways to make this language explicit with young learners so they are familiar with the different contexts.

A Child's Mind . . .

Counting On

"How far apart are these two numbers?" is a question about the distance between two numbers, which is a measurement idea. Some children with experience using game boards to count from one space to the next will see how they can respond to this question by *counting on* from the smaller number to the larger number. However, often children make the mistake of counting the number they are on as they count to find the difference between two numbers. It's important to model for them that what is being counted is the *space* between the numbers, or the number of jumps it takes to get from one number to the next. It may be helpful to act out the counting by pretending you are taking finger steps on the number path to get from one number to the next. Students are exploring ideas here about subtraction, or finding the difference, by counting up to the total.

Introducing 15

Using an Open Number Line with Numerals 200 and 500

Suggested Grade Level

Grade 2

Materials

open number line labeled *200* and *500*

Related Game

- G-3 Mystery Number (Version 3)

Key Questions

- Why do you think this is called an open number line?
- Why do you think [number] is there on the number line?

Teaching Directions

1. Prepare an open number line per the materials list. Make sure the number line is large enough for everyone in class to see it.

Open number line labeled *200* and *500*.

2. Gather students on the rug so that everyone has a good view of the number line. Seat them close together to make listening to each other easier.

3. Draw students' attention to the number line. Remind students that it is called an *open* (or *empty*) *number line*. Ask, "Why do you think this is called an open number line?" Most likely, students will respond with the idea that some of the numbers are missing or that only two numbers are shown. They may also notice that the numbers begin at 200 and go to 500.

4. Now ask, "What do the little arrows by two hundred and five hundred mean?" The response might be that the numbers can get smaller going to the left and keep on going beyond 500 to the right.

5. Ask, "Who can point to where you think three hundred would be on this line?" Have a student volunteer indicate where they think 300 would be on the line. Ask, "Why do you think this?" Some students may approximate where 300 and 400 might be by partitioning the space into thirds. Before marking *300*, check with the rest of the students to see if they agree with the location

Technology Tip

Using an Interactive Whiteboard
An interactive whiteboard can also be used to display an open number line. Use the line option to draw a straight line, then use the pen to label each end.

Teaching Tip

Reviewing Vocabulary
Rather than assume that children know the meaning of words used in math, take a minute or two to be sure that everyone has a common understanding of the vocabulary you'll be using during a lesson.

Teaching Tip

Being Comfortable with Approximations
Do not give students the option of using measuring tools when locating numbers on the number line during this activity. In this instance, we want students to begin to feel comfortable with approximation, using what they know about the relative size of the numbers. Instead of counting squares, like on a number path, we want students to be able to move by imaginary jumps of squares along the number line.

for 300. If there are other suggestions, have students come up and point to the location and explain why they think 300 belongs there. Students might note that 350 is halfway between 200 and 500.

6. Ask, "Who can point to where you think four hundred would be on this line?" Have a student volunteer indicate where they think 400 would be on the line. Ask, "Why do you think this?" Before marking *400*, check with the rest of the students to see if they agree with the location for 400. If there are other suggestions, have students come up and point to the location and tell why they think 400 belongs there. Students will most likely note that the distance between 200 and 300 is about the same as the distance between 400 and 500.

7. Now ask, "Who can point to where you think two hundred ninety-six would be on this line?" Have a student volunteer indicate where they think 296 would be on the line. Ask, "Why do you think this?" Before marking *296*, check with the rest of the students to see if they agree with the location for 296. If there are other suggestions, have students come up and point to the location and tell why they think 296 belongs there. Students will most likely note that 296 is a little less than 300.

8. As a closing activity, ask, "If we wanted to write all the numbers on this line, how many would there be?" Provide time for students to think alone and then share with a shoulder partner.

9. Ask students to explain how they know how many numbers would be on the number line. As you call on volunteers to share with the whole class, encourage them to use the number line as they explain their thinking.

Math Matters!

How Many Numbers Would There Be on This Number Line?

This can be an interesting problem for students to solve using a zero-to-ten number line as a model for thinking. When we begin with zero, there are eleven numbers from zero to ten. A zero-to-twenty number line would have twenty-one numbers. So, a 200–300 number line would have 101 numbers, a 200–400 number line would have 201 numbers, and a 200–500 number line would have 301 numbers.

A Child's Mind . . .

The Challenge of Digits

Some children may think of each of the numbers on the number line as three numbers, but it is only one number made with three digits. For more of a challenge, have students determine how many digits they would need to write if they filled in all the numbers on the number line.

Using an Open Number Line with Numerals 700 and 1,000

Introducing 16

Related Game

- G-3 Mystery Number (Version 3)

Key Questions

- Why do you think this is called an open number line?
- Why do you think [number] is there on the number line?

Suggested Grade Level

Grade 2

Materials

open number line labeled *700* and *1,000*

Teaching Directions

1. Prepare an open number line per the materials list. Make sure the number line is large enough for everyone in class to see it.

Open number line labeled *700* and *1,000*.

2. Gather students on the rug so that everyone has a good view of the number line. Seat them close together to make listening to each other easier.

3. Draw students' attention to the number line. Remind students that it is called an *open* (or *empty*) *number line*. Ask, "Why do you think this is called an open number line?" Most likely, students will respond with the idea that some of the numbers are missing or that only two numbers are shown. They may also notice that the numbers begin at 700 and go to 1,000.

4. Now ask, "What do the little arrows by seven hundred and one thousand mean?" The response might be that the numbers can get smaller going to the left and keep on going beyond 1,000 to the right. Some students may be able to respond that the next number on the number line, if we extended it, would be 1,001.

5. Ask, "Who can point to where you think eight hundred would be on this line?" Have a student volunteer indicate where they think 800 would be on the line. Ask, "Why do you think this?" Some students may approximate where 800 and 900 might be by partitioning the space into thirds. Before marking *800*, check with the rest of the students to see if they agree with the location for 800. If there are other suggestions, have students come up and point to the location

Technology Tip

Using an Interactive Whiteboard

An interactive whiteboard can also be used to display an open number line. Use the line option to draw a straight line, then use the pen to label each end.

Teaching Tip

Reviewing Vocabulary

Rather than assume that children know the meaning of words used in math, take a minute or two to be sure that everyone has a common understanding of the vocabulary you'll be using during a lesson.

and tell why they think 800 belongs there. Students might note that 850 is halfway between 700 and 1,000.

6. Ask, "Who can point to where you think nine hundred would be on this line?" Have a student volunteer indicate where they think 900 would be on the line. Ask, "Why do you think this?" Before marking *900*, check with the rest of the students to see if they agree with the location for 900. If there are other suggestions, have students come up and point to the location and explain why they think 900 belongs there. Students will most likely note that the distance between 700 and 800 is about the same as the distance between 800 and 900.

7. Ask, "Who can point to where you think nine hundred fifty-one would be on this line?" Have a student volunteer indicate where they think 951 would be on the line. Ask, "Why do you think this?" Before marking *951*, check with the rest of the students to see if they agree with the location for 951. If there are other suggestions, have students come up and point to the location and explain why they think 951 belongs there. Students will most likely note that 951 is a little more than halfway between 900 and 1,000.

8. Now ask, "Who can point to where you think eight hundred ninety-one would be on this line?" Have a student volunteer indicate where they think 891 would be on the line. Ask, "Why do you think this?" Before marking *891*, check with the rest of the students to see if they agree with the location for 891. If there are other suggestions, have students come up and point to the location and explain why they think 891 belongs there. Students will most likely note that 891 is closer to 900, but not right next to it.

Teaching Tip

Being Comfortable with Approximations
Do not give students the option of using measuring tools when locating numbers on the number line during this activity. In this instance, we want students to begin to feel comfortable with approximation, using what they know about the relative size of the numbers. Instead of counting squares, like on a number path, we want students to be able to move by imaginary jumps of squares along the number line.

Math Matters!

How Many Numbers Would There Be on This Number Line?

This can be an interesting problem for students to solve using a zero-to-ten number line as a model for thinking. When we begin with zero, there are eleven numbers. A zero-to-twenty number line would have twenty-one numbers. So, a 700-to-800 number line would have 101 numbers, a 700-to-900 number line would have 201 numbers, and a 700-to-1,000 number line would have 301 numbers.

A Child's Mind . . .

The Challenge of Digits

Some children may think of 700 as three numbers, but it is only one number made with three digits. For more of a challenge, have students determine how many digits they would need to write if they filled in all the numbers on the number line.

9. As a closing activity, ask, "If we wanted to write all the numbers on this line, how many would there be?" Provide time for students to think alone and then share with a shoulder partner.
10. Ask students to explain how they know how many numbers will be on the number line. As you call on volunteers to share with the whole class, encourage them to use the number line as they explain their thinking.

Lessons Using Number Paths and Number Lines

Why these lessons?

The lessons in this section are designed to introduce students to the number path and number line as tools for understanding our number system more completely, as well as to help students create a mental image of the number path and number line that they can use in subsequent lessons. Supporting students in learning to use a number path or number line as a tool for adding and subtracting is one of the goals of this resource, along with improving students' understanding of our base ten number system.

How much time do these lessons require?

Lessons in this section may take more than one class period. During the initial lesson, consider introducing the problem and giving students time to explore the problem. Begin the summary during the next math period. Saving the summary for a later math period, rather than working through all three stages in one period, gives students time to think about their exploration. In addition, students tend to become less restless and can talk longer about what they have learned when the summary takes place at the beginning of a period rather than at the end.

Some explorations may take two full math periods. At the end of the first day, facilitate a class discussion as a check-in with students. Ask questions to help them think about what they have been exploring (see the Teaching Tip below).

This discussion provides support for all students when they come back to complete the lesson activity on the second day. At the end of the second day, check in again with another class discussion. Have any new insights or discoveries been made?

If there are students who need five or ten minutes more than what has been allotted to complete exploration, reserve time for those students to work on the activity before you begin the summary.

Teaching Tip

Check-in Questions for Use at the End of the Exploration Stage of a Lesson

- What do you understand?
- What solution did you find?
- What difficulties are you having?
- What new insights or discoveries did you make?

What is my role as teacher during these lessons?

At every stage of the lesson, your role is an active one. During the lesson introduction, give directions clearly and concisely. We want students to understand the activity without giving away the answer.

As students work through the lesson, use the key questions (included in the "Key Questions" section in each lesson) to probe students' thinking and to extend their understanding. When students are struggling with a concept, work with them by asking questions and referring back to the introduction. Refrain from telling them the answer.

At the end of the lesson, facilitate a discussion on the mathematics goals of the lesson. This discussion might include introducing or revisiting key vocabulary, looking for patterns, and making generalizations. To begin the summary, you might find it helpful to allow students to talk to their partners before you begin a whole-class discussion. What discoveries did they make during the lesson? What did they learn during their explorations? Starting with partner talk allows students to feel the value of being listened to while practicing what they may say during the whole-class discussion.

After partner talk, call on a few students to share their thinking or their work. (Note that the goal is not to call on every student to come to the front of the room to share their thinking, simply because of the amount of time involved to complete this activity.) As students share their thoughts, record their thinking where everyone can see it. This helps students make connections between mathematical thinking and symbolic representation. Repeated opportunities to see the symbolic representations eventually help students begin to use representations on their own.

Lesson 1

Building a Number Path

Time

15–20 minutes, three to five times during the course of a week; this lesson may be repeated or extended in subsequent weeks

Materials

strip of paper cash register tape (10 feet or longer)

paper clips, 20 per pair of students

Building the Number Path Number Cards (Reproducible 1), 1 deck (made from 1 copy of the reproducible)

glue stick

Extension

strip of paper cash register tape (10 feet or longer), 1 per pair of students

paper clips, 20 per pair of students

Building the Number Path Number Cards (Reproducible 1), 1 deck (made from 1 copy of the reproducible) per pair of students

glue stick

Overview

During this whole-class lesson, the teacher displays three number cards at a time and asks students to arrange the numbered squares in order from least to greatest. The teacher begins with three consecutive numbers and can then choose to work with consecutive or nonconsecutive numbers, depending on the skills of the students. Next, students are asked to create a number path using number cards 1 through 20. The arrangements are eventually saved on a length of cash register tape for number games and activities. Each pair of students or table group can eventually have their own number path for future lessons and games.

Typically, young children can say the counting numbers in order, but they may not be able to read all the numbers or arrange them in order, especially if they are not consecutive. This lesson challenges students by requiring them to create a number path versus reading an already prepared one. The lesson is most appropriate for kindergarteners.

Related Game

- G-1 *Mystery Number* (Version 1)

Key Questions

- How would you put these numbers in order?
- How do you know the number belongs there?

Teaching Directions

1. Before beginning the lesson, roll out a ten-foot length of paper cash register tape and attach it horizontally to the wall at student eye level.
2. Gather students on the rug so that everyone has a good view of the strip of paper cash register tape. Seat them close together to make listening to each other easier.
3. Hold up three number cards (use cards numbered 1, 2, and 3 at first). Ask, "Who would like to come up and hold one of these numbers?"
4. Call on three students, giving one number card to each student, and tell them to face the class and put the number card under their chin.
5. Tell the class, "Think quietly to yourself for a moment about how we should arrange the numbers in order from least to greatest." Give students some time to think. Then, ask, "How should the students stand up here in front of the class so the numbers they are holding are in order? Remember to explain your thinking."
6. Call on students to tell the volunteers where to stand so they are arranged in order from least to greatest. Ask the rest of the students to give a thumbs-up signal if they agree with the order as shown.

Examples of Student Thinking

Jose: "When you count, you hear 'One, two, three.' So you hear that when you count these."

Martin: "You could look on the hundreds chart or a calendar for the numbers."

A Child's Mind . . .

Providing Time to Think

When we give students the message that we are interested in their thinking, we need to ensure that we allow time for them to think. If some students blurt out answers right away, other students do not have their chance to think. Sometimes teachers use a gesture to suggest *think time* by pointing to their head and nodding quietly. Being able to think quietly before responding to a question takes practice and reminders. Eventually students will appreciate the chance to think before answering a question.

Teaching Tip

Silent Signals

Young children always seem to want to share what they know. Teaching them a way to use a silent signal such as thumbs-up (or thumbs-down or in the middle) lets them show what they know without extra comments or conversation. This is also a way for the teacher to gather formative information about whether the students are making sense of the lesson. Some teachers have students pat their head if they want to agree with what someone else has said, to show "That was what I was thinking." Other teachers use the American Sign Language expression for "I agree with you," as seen in videos in the *Number Talks: Whole Number Computation* resource (for example, see Video Clip 2.1 at www.mathsolutions.com/NTWNC21 or at https://www.signingsavvy.com/sign/ME+TOO).

American Sign Language expression for "I agree with you."

Differentiating Your Instruction

Choosing Numbers

As you call students forward to participate in the activity, be mindful of what you know about their counting and number knowledge. For example, if you know that some students may not be able to arrange nonconsecutive numbers in order from least to greatest, you might choose two numbers that are consecutive, and the third number may be two more than the second number.

Math Talk

Explaining Their Thinking

Create a classroom where students learn from one another, not just from the teacher. Students benefit from explaining their thinking as well as listening to how classmates explain their ideas. Sometimes one student may have the same idea but may use a different way to explain his or her thinking. A student who may have been confused at first may better understand what the second speaker has said using different words.

7. Collect the three cards and repeat Steps 3 through 6 with the next three consecutive numbers: 4, 5, and 6. In this case, students will be *counting on* from a number other than one. This is different from beginning at one each time.

Examples of Student Thinking

Eva: "I know that it goes three, four, five, six."

Monica: "I used to be four, then I was five."

8. Next, select three nonconsecutive number cards such as 4, 6, and 7. Call on three more student volunteers, hand out the cards in a random order, and challenge the students to arrange themselves in order from least to greatest, without help from the rest of the class. Tell students that these numbers may not be in consecutive order, but they can still be arranged from least to greatest.

9. Ask some volunteers to confirm the resulting arrangement is correct and explain how they know this is so.

Examples of Student Thinking

Ramona: "Four comes before the five that is missing, and then seven is after the six."

Alicia: "If we had all the numbers it would go four, five, six, seven."

10. Repeat Steps 8 and 9 with three or four other numbers that are nonconsecutive, depending on how well the first group of students was able to do the task.

Examples of Student Thinking for How to Arrange Cards 4, 6, and 8

Erin: "Four comes before six, and eight comes after six and seven."

Arturo: "You don't have the five or the seven, but you still put them from little to bigger."

Ramona: "Eight and six is older than four."

11. Continue with three or four different numbers until every child has had a chance to be in front of the class or if the students seem ready for a greater challenge. Each time, ask students to explain how they know the number order is correct.
12. Now turn to the strip of paper cash register tape that you stuck on the wall during Step 1. Ask ten students to take a paper clip and one of the number cards 1 through 10. Explain, "I want you to clip these number cards on the paper cash register tape in order from least to greatest beginning with one."
13. When all the numbers are clipped to the paper, check with the class to confirm the order is correct.
14. Challenge the rest of the students to use the numbers 11 through 20 and do the same thing.

Teaching Tip

Spacing of Numbers

With student-made number paths, the spacing between the numbers may not be exact. Try to model putting all the (equal size) number squares right next to one another so the spacing is roughly equal. When constructed in this manner, this tool shows the magnitude of a given number in relation to other numbers. This helps children visualize the size as well as the relationship of one number to another.

A Child's Mind . . .

When to End the Lesson

Watch students carefully for signs of restlessness and fatigue. Because this lesson is intended for the beginning of the year, students will have limited ability to sit still while just a few students participate at a time. When students become squirmy, move to a different activity, then return to this one later. Or, you can have a few students remain to add more number squares to the paper strip while others move on. This lesson can be broken down easily into several sessions with no loss of continuity. Additional number path strips can become tools for small-group activities, so save partially completed strips and use them another day.

Differentiating Your Instruction

Creating Accessibility for All Students

In kindergarten, think about your students and their previous experience and comfort with ordering numbers. Young students will be challenged at first with numbers from 1 through 5 and eventually 1 through 10. Also, kindergarteners will be more successful with this lesson when they work in small groups, and they might use numbers 1 through 20 later in the school year. If you would like to create a challenge, numbers 21 through 50 will be needed. You can write the numbers on small paper squares with a marker, or students can do this themselves.

First graders could work with numbers 1 through 50 and move on to number paths that go from 100 to 150, 150 to 200, 200 to 250, and so forth. In second grade, the numbers could approach 1,000.

Extend Their Learning!

Kindergarten

On another day, after you have made a number path to 20 with the whole class, students could work in pairs to build their own number path to 10 or 20. They can paste the numbers on the number path or use paper clips. You can continue with longer paper strips, using numbers to 50 or to 100 if students seem ready to do this.

First and Second Grade

First and second graders would benefit from making number paths with chunks of numbers such as 41 through 50 or 15 through 30 or 120 through 140. These chunks do not start with one, so sorting out the number cards and deciding where to begin provides a challenge for older students. The paper number paths that students make may be used for class games and problem solving. (See Lesson 2, *Counting Cups on a Number Path* and Game 1, *Mystery Number*, in this resource.)

Counting Cups on a Number Path

Lesson 2

Overview

During this activity, students count a predetermined number of items, place them in a cup with the corresponding number, and place the cup on the number path where it belongs. Students practice making sense of the meaning of vocabulary words that support number relationships.

Students look at the collection of cups thus displayed on the number path and respond to questions involving language such as: *more, less, how many, one more, two more,* and *equal.* This lesson is most appropriate for kindergarteners and first graders.

Related Lesson

- L-10 Gathering and Estimating Data

Key Questions

- *How many* cups are on the number path?
- Which cup holds the *most* number of counters?
- Which cup holds the *least* number of counters?
- Which cups hold *more than* five counters?
- Which cups hold *less than* five counters?
- Where should we put this cup if we put *two more* counters in the cup?
- How do we know the number of counters is *equal to* the number on the number path?

Time

15–20 minutes; this activity is best completed during small-group rotations

Materials

number path with the numbers 1–20 written on a sentence strip or paper cash register tape (place the numbers about two to three inches apart so there is room for a plastic or paper cup to sit on each number)

clear plastic cups, 6–10, each with a number in the 1–20 range written on the side

counters (enough to fill all cups with your chosen numbers)

Extension

number path with the numbers 1–20 written on a sentence strip or paper cash register tape (place the numbers about two to three inches apart so there is room for a plastic or paper cup to sit on each number), per pair of students

clear plastic cups, 6–10, each with a number in the 1–20 range written on the side, per pair of students

counters (enough to fill all cups with your chosen numbers), per pair of students

Teaching Directions

1. Prepare a number path per the materials list. Make sure it is large enough so everyone can see it. Place it flat in the middle of the demonstration space. Place the cups and counters next to it.

1	2	3	4	5	6	7	8	9	10	11	12	13	14	15	16	17	18	19	20

2. Gather students around the demonstration space so that everyone has a good view of the number path. Seat them close together to make listening to each other easier.
3. Show students a plastic cup and ask them to read the number on the side of the cup.
4. Ask, "How many counters should we put in this cup?" If students do not respond with the same number that is written on the cup, explain to students that the number on the cup is the label that tells how many counters should go in the cup.
5. Have a student volunteer put the matching number of counters in the cup. Support the student in making an accurate count by counting out loud as they put the items into the cup. Encourage the rest of the students to join in.
6. Now ask the student volunteer to place their cup on the number path where it belongs. Ask the student to explain how they know this number is equal to the number of counters in the cup.

Examples of Student Thinking

YuYu: "This cup goes here. It's a five."

Ramon: "The number on the cup is the same as the number on the number path."

Bodie: "I just know my numbers."

Roberto: "They match."

Teaching Tip

What Range of Numbers Should Be on Your Number Path?

The directions for this lesson assume the use of a 1–20 number path. Determine whether this range is appropriate for your students. Be mindful of what you know about your students' counting and number knowledge. For instance, kindergarten students will likely be more comfortable with a number path to 10 at the beginning of the year and could work with number paths to 20 later in the year.

Differentiating Your Instruction

Creating Accessibility for All Students

Before beginning this lesson, think about your students and their previous experience and comfort with counting objects. You may need to provide number cups with smaller numbers at first and then have additional challenge cups for students who seem to find the task easy. It is interesting to provide a range of choices for students. First graders usually choose numbers at their own comfort level and, with more experience, venture into using numbers they find challenging, such as those more than twenty-nine.

7. Continue with additional cups, letting students choose from the remaining cups. Each time, ask the student to explain how they know where the cup belongs. Students benefit from explaining their thinking as well as listening to how classmates explain their ideas.

8. When all the cups have been placed on the number path, engage the students in responding to the key questions (page 77). (In the following examples, the teacher used beans for counters.)

Examples of Student Thinking

Teacher: "*How many* cups are on our number path?"

Sky: "There's nine cups."

Silvia: "You count them one, two, three, four, five, like that."

Teacher: "Which cup holds the *most* number of beans?"

Bodie: "This one, 'cause it's the biggest number."

Roberto: "The one with nine."

Teacher: "Which cup holds the *least* number of beans?"

YuYu: "One."

Sky: "It only has one bean."

Silvia: "Least is the smallest."

Teacher: "Which cups hold *more than* five beans?"

YuYu: "All these over here."

Nicole: "Six, seven, and nine."

Teacher: "Which cups hold *less than* five beans?"

Manuel: "One and three."

Teaching Tip

Pacing

Begin with a cup that is labeled with a number less than 10. Consider asking for volunteers at first who have demonstrated they can count accurately to the target number. If a student is not accurate, ask them to recount again with you.

Math Matters!

Two-Digit Numbers

Often, young children have difficulty reading two-digit numbers and they transpose the digits, saying ninety-one for nineteen or seventy-two for twenty-seven, for example. If necessary, read the number correctly for them, but do not stop at this time to explain the place-value concepts. Make a note of which students are confused about this and provide additional experiences at a later time to strengthen their skill. With more exposure to larger numbers and the patterns found in the place-value system, these students begin to read two-digit numbers correctly.

Teaching Tip

Supporting English Language Learners

Make sure you use words such as *least* and *greatest* in a context that makes sense to your students. Conversations using *more* and *less* also support students in developing understanding of the meaning of numerical order. Posters illustrating more and less can serve as important reference tools for second language learners in particular. An example of this type of poster is provided as Reproducible 2, Greater Than, Less Than, and Equal To Reference Chart.

Examples of contextual conversations for *more* and *less*:

"Are there more boys or more girls in this table group?"

"I think there are less students absent today than yesterday."

"We can fit two more students at this table."

"You can take at least three pretzels, but not more than five."

"He has the most cards."

"She has the least cards."

Teacher: "Where should we put this cup if we put *two more* beans in the cup?"

Bodie: "I don't get it."

[**Note:** It is much easier to think of *one more* than *two more* beans, so this question probably should be acted out.]

Teacher: "How many would we have if we put *one more* bean in the cup? Let's do it. Now how many?"

Teacher: "How do we know the number of beans *is equal to* the number on the number path?"

Nicole: "We have to change the number on the cup if we put more beans in the cup."

Students Playing Counting Cups on a Number Path

Extend Their Learning!

Partner Activity

After your students understand what they need to do when presented with the counters, labeled cups, and a number path, this activity can be used later on during choice time or as a learning station in math workshop. Determine the appropriate range of numbers for your student pairs and include some nonsequential numbers on the cups so that not all numbers in the range are represented. Students can fill in the blank numbers by making their own cups and matching contents, in addition to using the premade cups. (This additional challenge makes the task easier to check for accuracy.)

At-Home Activity

Give each student a strip of paper with a number path already on it and five to ten cups to take home for reinforcement and practice. I have found that parents appreciate the simplicity of this activity. They report their child enjoyed finding things to count and display at home.

Teacher Reflection

Students Exploring Independently

When my students knew what to do with the counting cups and the number path, I put blank strips of paper, some cups, markers, and counters in a bin and they set to work making their own number path, labeling their own cups, and counting the quantities they determined when they labeled the cups. Many students used their favorite numbers, such as their mom's age or the number of people in their family, for landmark numbers on their number path. It was interesting to see them so engaged in setting up their personal number path and matching the quantities to the numbers they selected.

Lesson 3

Building an Open Number Path

Time

15–20 minutes, three to five times during the course of a week

Materials

strip of paper cash register tape (about 10 feet in length)

3-inch-by-3-inch red paper squares, 20

3-inch-by-3-inch yellow paper squares, 15

glue stick

Twenty Hungry Piggies by Trudy Harris (or another favorite counting book)

pointer

Ordinal Number Reference Chart (Reproducible 3)

Extension

Dot Cards (Reproducible 4)

Differentiating Your Instruction

Creating Accessibility for All Students

Before beginning this lesson, think about your students and their previous experience with other number paths. Also, pay attention to the comfort your students have with number order.

Overview

In this lesson, students are introduced to ordinal number names by lining up and counting off according to ordinal number names. They then use a teacher-prepared number path of thirty-five alternating groups of five red and five yellow blank paper squares to find the landmark numbers of multiples of fives and tens. Students use ordinal and cardinal number names to describe the paper squares. It is important in this lesson to support this language development as well as the conceptual development of landmark numbers. Because the alternating color arrangement does not have numbers, we think of it as an open number path. Later on, in Lesson 4, *Jumping by Ones and Tens*; Lesson 5, *Jumping by Ones, Fives, Tens, and Twenty*; Lesson 6, *Solving Story Problems Involving Missing Numbers at the End*; Lesson 7, *Solving Story Problems Involving Missing Numbers in the Middle*; Lesson 8, *Solving Story Problems Involving Missing Numbers at the Beginning*; and Lesson 9, *Solving Comparison Problems*; students draw their own open number paths with landmark numbers rather than writing each number. When children can construct an image mentally of tens and fives as well as ones, they will have friendly numbers at their disposal that they can manipulate in a way that makes sense to them.

Related Games

- G-1 Mystery Number (Version 1)
- G-2 Mystery Number (Version 2)
- G-3 Mystery Number (Version 3)
- G-4 Race to 50
- G-7 Hot Lava Bridge: Forward and Back, More and Less
- G-10 Adding Nines, Tens, and Elevens

In these games the red and yellow open number path can be used instead of a number path with all the numbers. In Lesson 10, *Gathering and Estimating Data*, the open number path is used to record estimates.

Key Questions

- Which square is fifth? Tenth?
- What color is the fifteenth square? How do you know this without counting?
- Which squares are always red?

Literature Connection

20 Hungry Piggies

The book *20 Hungry Piggies* by Trudy Harris builds off the original nursery rhyme in which the fifth piggy goes "wee wee wee all the way home"—only he does not! Instead, more pigs become involved and gather gleefully for a picnic. What they soon realize is that the infamous wolf straight out of *The Three Little Pigs* has also decided to "picnic." The book is filled with personality and counting fun. Students enjoy the humor and begin to make connections to ordinal numbers up to twentieth.

Teaching Directions

1. Before beginning the lesson, prepare the open number path. To do so, roll out a ten-foot length of paper cash register tape and glue alternating groups of five red and five yellow blank paper squares, until there are thirty-five squares in total. Affix this red and yellow open number path horizontally to the wall at student eye level.
2. Gather students on the rug so that everyone has a good view of the number path. Seat them close together to make listening to each other easier.
3. If you have the children's book *20 Hungry Piggies* by Trudy Harris, set the stage for counting by reading it to your students. (If you do not have this book, choose one from your collection that supports counting.)
4. Introduce the vocabulary of ordinal numbers using the Ordinal Number Reference Chart (Reproducible 3).
5. Ask ten students to line up and count off in order, one by one, using ordinal numbers (first, second, third, and so on) to review the language. Pay special attention to the ending sounds of the words. Have ten more students line up and count off. Do this as needed to ensure every student gets a turn.
6. Now draw students' attention to the number path. Point to the first square and connect it to the number 1 as well as the *first* square. Do the same with the second square.

7. Now say, "Think about how many squares there might be on the paper strip. How do you know?"

8. Call on students to tell the class what they notice about the paper strip. Some students may count all the squares, beginning with the first one, whereas others may be able to count by fives.

9. Ask a student volunteer to use the pointer to find the number 5 square. Explain that this square is called the *fifth*.

10. Instruct the students to count "One, two, three, four, five" to the fifth square. Use the pointer, beginning with the first square, as they count. Write *fifth* above the fifth square to emphasize the name of this square.

11. Ask a student volunteer to choose a number on the Ordinal Number Reference Chart and, using the pointer, find that number square on the open number path. Have the class count up to this number, beginning with the first square while pointing at each square. Write the ordinal number name where everyone can see it to emphasize the spelling and pronunciation of this name.

12. Ask another student to choose a number on the Ordinal Number Reference Chart and find the matching square on the number path. Have the students count up to this number, beginning with the first square while you point at each square.

13. Now ask students to locate the tenth square with their eyes. Choose a student to point it out with the pointer. Count up to the tenth square from the first square. Ask students to explain how they were able to find the tenth square with their eyes.

Teaching Tip

Choosing Numbers

As you call students forward to point to different numbers, begin with numbers between 1 and 10, then move on to multiples of five, taking advantage of the groups of five for landmarks. In this lesson you help students focus on the landmark numbers of 10, 20, and 30 as well as multiples of five. Often, teachers help students think about these landmark numbers by referring to them as "friendly" numbers.

A Child's Mind . . .

Locating with Your Eyes

When children scan the number path to "locate with their eyes" instead of counting up to the number, they are beginning to think of the unit of five or ten as one group. Being able to find these landmark numbers with their eyes develops with more experience and is supported by the counting you do orally with them to determine whether they are actually focusing on the number in question.

14. Next, ask students to locate the twentieth square with their eyes. Choose another student to point it out with a pointer. Count up to the twentieth square with all the students.

15. Ask students to explain how the number path was constructed. Ask, "If we were to make this path longer, what would we do next to make it longer?"

16. Ask, "How many squares would there be if five more squares were added to the path? What color would come next if we were to extend the path?"

17. After students become familiar with finding the matching number on the number path, this activity can become a partner or small-group activity used during choice time or as a learning station in math workshop.

A Child's Mind . . .

When to End the Lesson

Watch the class carefully for signs of restlessness and fatigue. Because this lesson is intended for the beginning of the year, students will have limited ability to sit still while just a few students at a time participate. When students become squirmy, move to a different activity, then return to this lesson later. Or, you can have a few students remain to add more color paper squares to the paper strip while others move on. This lesson can be broken down easily into several sessions with no loss of continuity.

Math Matters!

Patterns on the Open Number Path

On the open number path used in this lesson, there are five repeating red squares followed by five repeating yellow squares (it could be called an AAAAABBBBB pattern). Note that this is a repeating pattern, whereas other patterns may grow, such as in this visual:

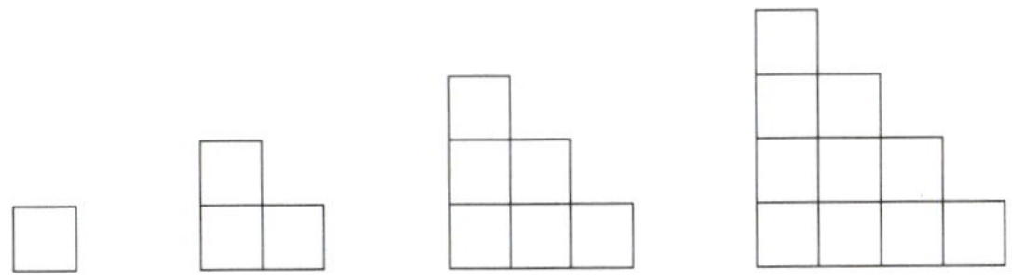

Growing pattern.

What is important about this open number path is that it is a tool that shows groups of fives that are contained in a group of ten. Each group of ten is composed of five plus five. Color helps students see the part–part relationship. We could make a strip of three red, three yellow, three red, three yellow, and so on, but that arrangement is not useful in helping students make sense of base ten place value.

Extend Their Learning!

Using Dot Cards

1. Show students a dot card (Reproducible 4) with an arrangement of dots in groups of fives or tens and possibly one or two extras. (Note that dot cards can be made easily. One way is to put garage-sale dot stickers on paper plates.)
2. Ask students to find the square on the number path that matches the number of dots on the card.
3. Watch to determine whether students are able to identify the number of dots quickly by noticing the groups of five or ten arranged in frames or dice patterns.

Continuing Past Twenty

After you have made an open number path to twenty, you can continue with paper squares to 50 or 100. The number path can be used for class games and problem solving (see "Related Games," page 83). When children play games on an open number path that stretches to 50 or 100, they are making sense of the number relationships by looking at the position of the paper square without "reading" the number label. The magnitude of the number is represented by how far along the strip of paper the square is located.

Math Matters!

Subitizing

Subitizing is the ability to glance at a group of objects and see quickly how many objects there are without counting them one by one. When children notice there are five squares of one color in a row on an open number path, they can see that the next group also has five squares (see *Math Matters: Understanding the Math You Teach, Grades K–8,* 2nd edition, by Suzanne H. Chapin and Art Johnson [Math Solutions, 2006]). For more lessons on subitizing, see Routine 1, *Look, Quick!*; Routine 2, *Make the Number*; and Routine 3, *Say the Two-Digit Number* from *It Makes Sense! Using Ten-Frames to Build Number Sense* by Melissa Conklin (Math Solutions, 2010).

Jumping by Ones and Tens

Lesson 4

Overview

In this lesson, students move a marker from the center of a number path according to the results of spinning a spinner and drawing a card. The goal is to get to either end of the number path.

Students begin with jumping from one place to another by small jumps of one to five spaces each. Students will end up moving back and forth; however, there won't be very much movement at first. When the idea of being able to make jumps of ten is introduced, students make predictions of what will happen when larger jumps are made. The students then try using this new option.

This is a cooperative activity instead of a competitive game. The goal is for the players to move to the end of the number path (or off it). This lesson is most appropriate for kindergarteners.

Related Games

Games using the red and yellow open (numberless) number path or the numbered number path:

- G-4 Race to 50
- G-7 Hot Lava Bridge: Forward and Back, More and Less
- G-11 Get to the Target (Version 1)

Time

15–20 minutes, three to five times during the course of a week

Materials

number path labeled *1–20* or open number path created with 20 paper squares alternating five red and 5 yellow

Amanda Bean's Amazing Dream by Cindy Neuswchwander (or another favorite counting book)

clothespin or large paper clip, 1 (to be used as a game marker)

Jumping by Ones and Tens Number Cards (Reproducible 5), 1 set (made from 1 copy of the reproducible)

paper lunch sack, 1 (to hold the number cards)

More/Less Spinner (Reproducible 6)

large paper clip (to spin the spinner)

pencil (to spin the spinner)

Extension

number path labeled *1–20*, 1 per team or pair of students (or number paths constructed in L-1 or sentence strips for making number paths)

clothespin or large paper clip, 1 per team or pair of students (to be used as a game marker; if using with teams, make sure markers are different colors)

Jumping by Ones and Tens Number Cards (Reproducible 5), 1 set per team or pair of students (made from 1 copy of the reproducible)

paper lunch sack, 1 per team or pair of students (to hold the number cards)

More/Less Spinner (Reproducible 6), 1 per team or pair of students

large paper clip (to spin the spinner), 1 per team or pair of students

pencil (to spin the spinner), 1 per team or pair of students

Literature Connection

Amanda Bean's Amazing Dream
In the book, *Amanda Bean's Amazing Dream*, Amanda loves to count things, but grapples with how to count them more efficiently. She falls asleep dreaming of a tranquil bike ride in the country, which quickly turns into counting chaos, ultimately convincing Amanda to learn new ways of counting. The book is filled with humorously illustrated objects to count.

Key Questions

- How do you know how far to jump without counting each place?
- How do you know you've landed in the correct place?
- How far from 10 are you?
- How far are you from the end? How do you know?

Teaching Directions

1. Prepare a number path per the materials list. In this lesson you can use the number path labeled 1 through 20 or the open (numberless) number path with twenty paper squares alternating five red and five yellow squares as in Lesson 3, *Building an Open Number Path*. Make sure the number path is large enough for everyone in class to see it. Display it on the wall at student eye level.
2. Prepare the number cards. Remove the number 10 and 20 cards from the set and place the remaining cards in a paper sack nearby. Also, set a game marker (clothespin or large paper clip) and the More/Less Spinner nearby.
3. Gather students on the rug so that everyone has a good view of the number path. Seat them close together to make listening to each other easier.
4. If you have the book *Amanda Bean's Amazing Dream* by Cindy Neuschwander, set the stage for counting by reading it aloud to the students. (If you do not have this book, choose one from your collection that supports counting.)

5. Draw students' attention to the number path and explain the activity. Say, "Today we are going to choose a number card that tells how many spaces we have to move. We spin the spinner to find out whether we need to move forward (more) or back (less). We'll begin in the middle. Our goal is to see if we can move to one of the ends or off this number path."

6. Ask students, "Where should we place the marker if we want it to be in the middle of the number path?" When students indicate the middle of the path, ask, "How do you know this is the middle?"

Examples of Student Thinking

Jenny: "You could measure with your arms like this and see that it's in the middle."

Camden: "The middle is ten because we can count to ten and we can count ten more to twenty."

Monica: "We could go in from the ends to get to the middle." (When Monica suggested this, she and the teacher modeled moving one step at a time in unison until they reached the middle of the path.)

7. Place the marker on the number 10 by clipping it over the number path. Call on a student volunteer to draw a Jumping by Ones and Tens Number Cards (Reproducible 5) from the paper sack. Ask another student volunteer to spin the More/Less Spinner (Reproducible 6). Model how to move the marker based on these directions.

Teaching Tip

More or Less

It might be helpful to draw arrows above the number path to indicate directions for more and less. Young children are just learning the meanings of these words as they apply to math.

← less more →

Assessment Opportunity

Using a Number Path to 30

On another day, use a number path that reaches to the number 30 and ask where the middle is for this number path. This question can be an informative assessment question to pose after you have established the meaning of a middle number.

Example of How to Model the Directions

Teacher: "Because Jose's spinner result showed *more*, we know we're going to move in this direction [gesture to the right]. Because Shirin drew the number four card, we'll jump four spaces, starting from ten in the middle: eleven, twelve, thirteen, fourteen. We are now six away from the end, so maybe in the next few turns we'll be off the number path."

(Place the number card back in the paper sack and begin the next round.)

8. Ask for two new student volunteers to draw a number card from the paper sack and spin the More/Less Spinner.
9. This time, ask students to think-pair-share where the marker should be moved as a result of the new card and spin.
10. Then have the volunteers move the marker from 14 to the new location. Have students confirm that the marker is in the correct location.
11. Ask, "Now how far are we from the end? How do you know?"
12. Continue the activity until the marker moves off the number path, either beyond 20 or to less than 1. Establish with students that this activity is only using small number jumps. Challenge students to think about what might happen if we put some number 10s in the sack.
13. Have students think-pair-share their ideas about what might happen if *more* and 10 were the moves.

Math Talk

Think-Pair-Share

When teachers ask a question to the whole class, often one student responds. However, when teachers use the think-pair-share strategy, all students get a chance to talk about the question. This can be a chance for students to rehearse what they may want to say to the whole class with a partner or an opportunity for them to predict what will happen in a safe setting. With experience, students become comfortable with the think-pair-share strategy and appreciate the chance to express their ideas before voicing them to the whole class.

Discussion Buddies

Often, students don't know who to turn to when teachers ask them to talk to a partner. To save time and alleviate any issues that might arise, instruct students to share their thinking specifically with a *shoulder partner*.

Examples of Student Thinking

Arturo: "We might get a ten on the first turn and be done."

Sarah: "Maybe we won't get a ten and we could get two fives instead."

14. Say: "We're going review how to add tens to a number so we can see how this might work. Let's count by tens to one hundred: ten, twenty, thirty, forty, fifty, sixty, seventy, eighty, ninety, one hundred."
15. As students count, write the numbers on the board in a vertical column:

    10

    20

    30

    40

    and so on. Ask, "What do you notice when we count by tens?"
16. Now have students count by tens, only starting at four. Say, "Now we are going to count by tens, but instead of starting at one we'll start at four." Begin a new column on the board. Ask, "What is ten more than four?" Count forward ten on the number path to 14. Say, "We noticed a pattern with the numbers when we counted by tens. Let's keep on counting by tens from four and see what patterns we notice." Continue, writing the numbers as students say them:

    4

    14

    24

    34

    44

    and so on.
17. Now draw students' attention to both columns. Ask students whether they notice any patterns in each column.

A Child's Mind . . .

Counting On

Watch to see whether students need to count one by one to arrive at the next number. Some children are confused about counting on and may count the number they are *on* rather than the *next* number on the number path. They need to begin with counting one more. Eventually, with more confidence, they can count two or three more. Board games during which students roll dice and move markers help them begin to make sense of how to count on from a specific number or place. Consider having these kinds of board games available for students who need more opportunities to practice counting on from a given space or number.

Teaching Tip

Moving Beyond the Number Path

In this activity, if *more* and 10 are the moves, a player at number 15 would move ten ahead and be beyond the end of the path. Similarly, if *less* and 10 are the moves, a player at number 5 would move ten back and be beyond the end of the path. Without spending time on the meaning of negative numbers at this time, explain that the activity board is just the number path from 1 to 20, and counting forward or backward can take a player off the number path and therefore end the activity.

Technology Tip

Using an Interactive Whiteboard

If you have a hundreds chart in the room or are using one on an interactive whiteboard, highlight the numbers as you count so the pattern becomes more apparent to students, thus helping them count by tens.

Assessment Opportunity

Choosing Number Paths for Lessons

Observe how comfortable your students are when they count by tens beginning with numbers other than ten. If they are able to notice patterns in the tens digits, then consider using a number path from 1 through 50 for this activity on another day. With 50 spaces on the number path, you can also include 20 as one of the number cards in the paper sack if your students seem to grasp what adding or subtracting twenty means.

If students are not comfortable counting by tens from any starting number, refer to and teach the lessons from *It Makes Sense! Using the Hundreds Chart to Build Number Sense* (Conklin & Sheffield 2012) and *It Makes Sense! Using Ten Frames to Build Number Sense* (Conklin 2010). At first, many students will count on their fingers to determine ten more, or need to refer to a hundreds chart as they count on ten more. With more experience, they will begin to notice the patterns in the ones and tens places.

Math Matters!

Unitizing

Eventually, we want students to think of a group of ten as ten jumps of one step each, as well as one jump of ten. When students count by tens, they are beginning to see these landmarks as important chunks *and* as individual numbers. When we move back and forth on the number path, we can move by units of five or units of ten as well as individual ones. The arrangement of five red and five yellow squares is a visual way to help students begin to think of the ten squares as a unit. With the open red and yellow number path, students begin to notice how jumps of ten in each direction land on the same color, in the same position of the group of five squares.

Examples of Student Thinking

Martin: "The numbers going down are more each time."

Jason: "The fours are the same all the way down."

Marsha: "It goes blank, one, two, three, four, five, six, seven, eight, nine, and the next number will be ten, I think."

18. Now say, "I think we're ready to make this more of a challenge." Add two number cards labeled *10* to the paper sack, shake it up, put the marker back at the middle of the number path, and call on two more volunteers to draw a number card and spin the More/Less Spinner.

19. Once again, ask student volunteers to tell the class what they will do on this turn. Ask students to place the marker on the number path accordingly and have the rest of the students confirm with a thumbs-up signal if they agree with the placement.

20. Before continuing with the next turn, ask students to predict what will happen if they get a 10 card with a spin of *more*. What might happen if they get a 10 card and a spin of *less*?

21. Continue class game until they move off the number path.

22. Challenge students to play this game again, but in small groups. Provide table groups with full sets of number cards, spinners, and markers. For the game board, give them the choice of using either a number path with numerals or the red and yellow open (numberless) number path. Ask students to pay attention to how many turns it takes before they move off the number path when they put the two 10 cards in their paper sack.

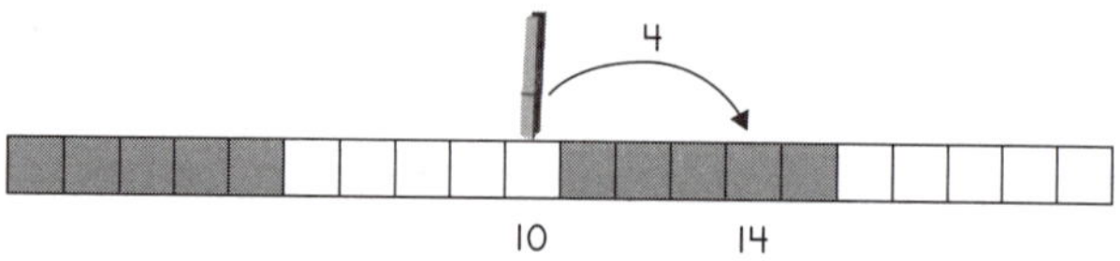

Shirin's move on the number path after drawing the number four card (and after Jose spun "more").

Extend Their Learning!

Moving to an Open Number Path

Use this activity first with a number path labeled *1* through *20*. On another day, play the same game with a red and yellow open (numberless) number path that may go up to the number 35. In this way, you will challenge your students to think about two groups of five as ten.

Competitive Class Game

After students are comfortable with the basic activity, consider playing it as a class in teams. Assign each team a colored game marker (clothespin or large paper clip), then explain that the two teams will take turns moving their marker according to the spin and draw of the cards. Before each turn, ask, "What do you hope to get on your spin? Explain to your team members why you think this is a good spin." Teams can then huddle to predict and share with each other what number and direction they want to get from their turn.

Math Talk

Making Predictions

When students talk about what might happen and explain why this is a good move, they are more engaged in the game and are making predictions about the results of an abstract number that could be chosen. This practice is also a formative assessment tool that you can use to determine whether the students are making sense of the consequences of the spins and draws.

Cooperative Partner Play

After students are familiar with the activity, they can play with a partner at their desks using tiles or paper clips as markers on their own number path. They can use the number paths they constructed in Lesson 1, *Building a Number Path*, or make their own using sentence strips. As students work, circulate and make observations about how they are determining where to move their markers. Are they counting one number at a time, or are they making jumps of ten?

Lesson 5 Jumping by Ones, Fives, Tens, and Twenty

Time

15–20 minutes, three to five times during the course of a week

Materials

strip of paper cash register tape (about 10 feet in length)

3-inch-by-3-inch red paper squares, 25

3-inch-by-3-inch yellow paper squares, 25

One Is a Snail, Ten Is a Crab: A Counting by Feet Book by April Pulley Sayre (or another favorite counting book)

clothespin or large paper clip, 1 (to be used as a game marker)

More/Less Spinner (Reproducible 6)

Jumping by Ones, Fives, Tens, and Twenty Spinner (Reproducible 7)

large paper clip (to spin the spinner)

pencil (to spin the spinner)

Overview

During this lesson for first graders (later during the school year) and second graders, students move a marker along a number path according to the results of two spinners. Students jump forward or backward one, five, ten, or twenty spaces. The goal is to jump until either end of the number path is reached. This lesson is built on Lesson 4, *Jumping by Ones and Tens*, during which jumping by tens is introduced. It is much more challenging for students to jump by fives. The red and yellow number path with color-coded groups of five supports students in adding and subtracting five as a chunk instead of counting on one by one.

Related Games

Games using the red and yellow open (numberless) number path or the numbered number path:

- G-4 Race to 50
- G-7 Hot Lava Bridge: Forward and Back, More and Less
- G-11 Get to the Target (Version 1)

Key Questions

- How do you know how far to jump without counting each place?
- How do you know you've landed on the correct place?
- How far are you from the end? How do you know?

Teaching Directions

1. Before beginning the lesson, prepare an open number path. To do so, roll out a ten-foot length of cash register paper tape and glue alternating groups of five red and five yellow paper squares, until there are fifty squares in total. Tape this red and yellow open number path horizontally to the wall at student eye level. Make sure the number path is large enough for everyone in class to see it. Display it on the wall at student eye level.
2. Place a game marker (clothespin or large paper clip) and the two spinners nearby.
3. Gather students on the rug so that everyone has a good view of the red and yellow open number path. Seat them close together to make listening to each other easier.

Extension

number path labeled *1–50*, 1 per team or pair of students (or number paths constructed in Lesson 1 but extended to 50 or sentence strips for making number paths)

clothespin or large paper clip, 1 per team or pair of students (to be used as a game marker; if using with teams, make sure markers are different colors)

More/Less Spinner (Reproducible 6)

Jumping by Ones, Fives, Tens, and Twenty Spinner (Reproducible 7)

large paper clip (to spin the spinner), 1 per team or pair of students

pencil (to spin the spinner), 1 per team or pair of students

Literature Connection

One Is a Snail, Ten Is a Crab

In the book, *One Is a Snail, Ten Is a Crab*, a cast of loveable critters offer their feet for counting—from crabs to dogs. Bright summertime illustrations support students in an informal exploration of how things can come in groups of twos, fives, and tens. Thinking about these groups supports thinking about what it means to count by them.

4. If you have the book *One Is a Snail, Ten Is a Crab* by April Pulley Sayre, set the stage for counting by reading it aloud to your students. (If you do not have this book, choose one from your collection that supports counting.)

5. Draw students' attention to the number path and explain the activity. Say, "Today we are going to continue jumping on the number path, but now we'll use these spinners to determine the size and direction of the jumps. What do you notice about the spinners?"

Examples of Student Thinking

Angela: "You can jump by one, five, ten, or twenty."

Dennis: "Sometimes you go more and sometimes you go less."

6. Continue, "We are going to begin in the middle of the number path. Our goal is to see if we can move off the number path."

7. Ask students, "Where should we place the marker if we want it to be in the middle of the number path?" After students answer, ask, "How do you know this is the middle?"

8. Place the marker on the 25th paper square by clipping it over the number path.

Examples of Student Thinking

Martin: "You go in ten from each end and you get to ten and forty, then twenty and thirty. Then you go to twenty-five in the middle."

Serena: "It's from one to fifty, so it's twenty-five steps."

Alicia: "Two people walk their fingers from the ends at the same time and they get to twenty-five in the middle."

A Child's Mind . . .

Thinking about Groups of Numbers

In this game, students first play using an open number path with two colors and no numerals. The game can later be played with the number path using numbers instead. On the open number path, the pattern of five red and five yellow squares supports students in thinking about counting by fives as well as by groups of ten. When students are familiar with the idea that a number such as seven is two more than five, they begin to see that seven plus five is two more than ten, and seven plus ten is two more than fifteen. Similarly, they also notice that nine is one less than ten and all the numbers with nine in the ones place are next to the multiples of tens.

9. Call on two student volunteers to spin the spinners. Model how to move the marker based on these directions. Say, "Because Marco's spinner resulted in more, I know we're going to move in this direction [gesture to the right]. Because Carla's spinner landed on twenty, we're going to move twenty steps forward." Ask, "How should we do this jump?"

Examples of Student Thinking

Jose: "We could jump two jumps of ten" [shows a jump to 35 and then to 45].

Martin: "We could just count up twenty more" [puts his finger on 25 and counts up to 45].

Teacher: "We landed on forty-five. Tell your shoulder partner what you want to get on the next spins if you want to get to the end of the number path."

10. Ask for two new volunteers to spin the spinners. Have students think-pair-share where the marker should be moved as a result of the new card and two spins.
11. Ask volunteers to move the marker, then confirm with all students that the marker is in the correct location.
12. Now ask, "How far are we now from the end? How do you know?"
13. Continue the game until the marker moves off the number path, either beyond the 50 or less than the 1.

Differentiating Your Instruction

Moving to a Number Path with Numerals
Use this activity first with an open red and yellow number path. On another day, play the same game with a number path labeled *1* through *50*. Eventually, students will see how the jumps of ten or twenty work and note the digit in the ones place is the same.

Assessment Opportunity

Using an Open Number Path to 100

Observe how comfortable your students are when they count up or down by fives and tens. If you feel they are ready to advance to using a number path of 0 and 100, to keep the game from becoming too long, create a spinner using larger numbers such as 30 and 40. Alternatively, use a number path that ranges from 50 and 100 so students can work with larger double-digit numbers, but still have a range of only fifty spaces on the number path.

Assessment Opportunity

Varying the Open Number Path

On another day, use an open number path labeled *0* and *60* and ask where the middle is for this number path. This question can be an informative assessment question to pose when you have established the meaning of a middle number. For another variation, consider using a partial number path from 20 to 50 instead of 0 and 50. Ask students to determine where the middle of this number path would be. Have them explain how they know this is the middle.

Math Talk

Think-Pair-Share

When teachers ask a question to the whole class, often one student responds. However, when teachers use the think-pair-share strategy, all students get a chance to talk about the question. This can be a chance for students to rehearse what they may want to say to the whole class with a partner, or a chance for them to predict what will happen in a safe setting. With experience, students become comfortable with the think-pair-share strategy and appreciate the opportunity to express their ideas before voicing them to the whole class.

Math Talk

Discussion Buddies

Often, students don't know who to turn to when teachers ask them to talk to a partner. To save time and alleviate any issues that might arise, instruct students to share their thinking specifically with a *shoulder partner*.

Extend Their Learning!

Competitive Class Game

After students are comfortable with the basic activity, consider playing it as a class in teams. Assign each team a colored game marker (clothespin or large paper clip), then explain that the two teams will take turns moving their marker according to the results of the two spinners. Before each turn, ask, "What do you hope to get on your spins? Explain to your team members why you think these are good spins." Teams can then huddle to predict and share with each other what number and direction they want to get from their spins.

Math Talk

Making Predictions

When students talk about what might happen and explain why this is a good move, they are more engaged in the game and are making predictions about the results of an abstract number that could be spun. This practice is also a formative assessment tool you can use to determine whether students are making sense of the consequences of the spins.

Cooperative Partner Play

After students are familiar with the activity, they can play with a partner at their desks using tiles or paper clips as markers on their own number path. They can use the number paths they constructed in other lessons or make their own using sentence strips. As students work, circulate and make observations about how they are determining where to move their markers. Are they counting one number at a time, or are they making jumps of five or ten?

Skip Counting

To help students with their counting and mental math for this lesson, consider skip counting as a class. Say to students, "Count by fives with me. I want to record the numbers to see if there is a pattern." Record the responses vertically so students are more able to notice the patterns.

5

10

15

20

25

30

35

40

45

50

and so on.

Ask, "What do you notice?"

Examples of Student Observations

Ramona: "The numbers are five and zero when you go down."

Marcos: "You have two numbers in the tens and two numbers in the twenties and then two numbers in the thirties, and like that."

Then say, "Now let's count from one, five more each time. I'm going to write what you say, and we'll look to see if there is a pattern when we count by fives starting with one." (Note: This will be much more challenging. Ask the class to count on orally with you five more from one using their fingers or using a hundreds chart to get to six. Then write 6 under 1.)

1

6

Now, ask the class to count on orally with you five more from the 6 using their fingers or a hundreds chart to get to 11. Then write 11 under the 6.

1

6

11

Then say, "Now let's count five more from eleven." (Use fingers or the hundreds chart.) Write 16 under the 11.

1

6

11

16

You can say, "Let's keep on counting and see how far we can go."

16

21

26

31

36

41

46

51

and so on. Now students have enough data to look for a pattern of how counting by fives from one repeats the digits in the ones place. Ask, "What do you notice?"

Examples of Student Observations

Carla: "It goes six, one, six, one on the way down."

Martin: "It doesn't have zero or fives like before."

Ask, "What happens if we begin with two and count five more each time?"

(Do this using the red and yellow open number path to count on to the next number this time instead of referring to the hundreds chart or using fingers.) Record the answers vertically.

2
7
12
17
22
27
32
37
42
47
52
57

Ask, "Now what do you notice?"

Examples of Student Observations

Eva: "Now you have two and seven on the way down."

Camden: "The numbers go down in order: one, one, two, two, three, three, four, four, five, five, ..."

Solving Story Problems Involving Missing Numbers at the End

Lesson 6

Overview

The group of story problems in this lesson are referred to as *result-unknown* (or *total-unknown*) problems. What is a result-unknown problem? Let's consider three examples:

1. Two frogs are sitting on a log. Three more frogs hop up to join them. How many frogs are now on the log? (Add to) 2 + 3 = ?
2. Five pencils are in a cup on the table. I took 2 of them. How many pencils are now in the cup? (Take from) 5 − 2 = ?
3. Three red marbles and 2 green marbles are in my pocket. How many marbles are in my pocket? (Put together/take apart when the total is unknown) 3 + 2 = ?

What do these three problems have in common? In result-unknown problems, the unknown number is at the end of the equation. This type of problem is most commonly addressed in school. The structure of these problems involves two or more quantities that are combined to obtain the sum (part-part-whole). After these ideas have been established, a total quantity is given and some of it is removed.

Time

15–20 minutes per problem

Materials

number path labeled *1–20*, 1 per pair of students

Story Problems: Solving Story Problems with Missing Numbers at the End (Reproducible 8), each problem on a separate sheet of paper, 1 set (4 problems) per student

counters (as needed to help students solve problems)

Related Lessons

- L-3 Building an Open Number Path
- L-4 Jumping by Ones and Tens
- L-7 Solving Story Problems Involving Missing Numbers in the Middle
- L-8 Solving Story Problems Involving Missing Numbers at the Beginning
- L-9 Solving Comparison Problems

Key Questions

- Who would like to explain what is happening in this story problem using your own words?
- What did we start with? What did we end up with?
- What happened?
- What do we want to know?

Teaching Directions

Problem One: *Seagulls*

1. Write the following story problem where everyone in the class can see it. Include blanks instead of numbers, as shown here.

   Problem One: Seagulls

   ___ seagulls are sitting on the lunch table during recess. ___ more join them when [name of student in the class] opens up a sandwich wrapper. How many seagulls are now sitting on the lunch table? What do you think happened?

2. Read the problem together with students. This can be done with the entire class or in small groups. The paper, with the problem, number path, and the counters can be made available to students after the students have had an opportunity to make sense of the problem.

3. Use the following statements to facilitate a productive discussion about the story problem:

   - "Tell the story problem in your own words."
   - "Share a different way to explain what is happening in the story."
   - "Tell your shoulder partner what this problem is all about."

Teaching Tip

Why Blanks for Numbers?
When introducing story problems, it's recommended to use blanks for the numbers at first, so students do not try to figure out an answer right away. Rather, you want them to focus on understanding what's happening.

4. Now insert numbers in the story problem as follows:

   Problem One: Seagulls

   Four seagulls are sitting on the lunch table during recess. Seven more join them when [name of student in the class] opens up a sandwich wrapper. How many seagulls are now sitting on the lunch table? What do you think happened?

5. Reread the story problem with students. Ask the following questions.
   - "Now that we know the number of seagulls in the problem, what happened at the beginning?"
   - "What do we want to find out?"
   - "Who has a way to explain what to do?"

6. Have students solve the problem. Suggest they use the number path to help solve the problem. You could ask, "Where should we start on the number path when [name of student] first comes to the lunch table?"

   Ask a student volunteer to explain how they solved the problem. Using the number path, the teacher (or the student sharing the strategy) acts out the problem, beginning at 4 and finger walks seven spaces to 11. Ask, "Who has another way to solve the problem?" Some students may add six and one instead of seven to show how to add up to the friendly number of ten and then add one more to get to eleven.

   Then draw an open number line, with arrows at both ends, on the board and say, "We can also use an open number line to solve this story problem." Mark a 4 near the left of the line. Say, "We can show the four seagulls being joined by seven more." Record what the student says using an open number line.

Technology Tip

Using an Interactive Whiteboard

If an interactive whiteboard is available, display the problem first with blanks for the numbers. Then, use the pen tool to write in the missing numbers after students have an understanding of what is happening in the problem.

Teaching Tip

Why These Numbers?

The numbers in this story problem have been chosen because the sum is greater than the ten fingers that students may try to use as counters in this problem. If the problem cannot be solved by counting with fingers, students begin to use addition strategies or counters to represent the numbers. These methods eventually allow children to think strategically about what happens when numbers are combined.

Technology Tip

Using an Interactive Whiteboard

If an interactive whiteboard is available, use the line option to draw a straight line, then use the pen to label each end with arrows and numbers.

Example of Student Thinking and Teacher Recording

Jenny: "I started with seven and added four more by counting. I ended up at eleven."

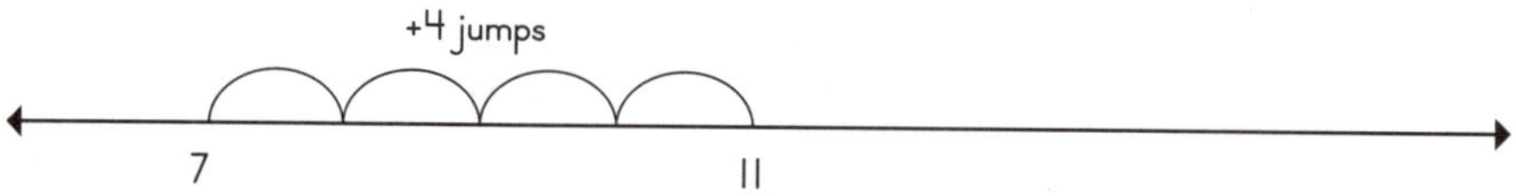

Recording Jenny's thinking on an open number line.

Note that Jenny decides to start with the larger number of seagulls and count on. Many students do not yet know that this is a more efficient way to count on.

> **Teaching Tip**
>
> **Recording Using a Mathematical Symbol**
> In this case, the teacher is the one who inserts the symbol (+) into the representation when the student talks about *seven more* seagulls coming to the table. You make the connection to the visual representation while the student explains the problem.

7. When translating the words students use to a teacher-constructed visual diagram, the teacher is the one who inserts the (+) into the representation. When the student talks about seven more seagulls coming to the table, the teacher writes (+ *7*).
8. Ask the student volunteer whether your recording (drawing) represents accurately what they said. If the student says it does not, ask the student to repeat their thinking as you change the recording accordingly.
9. Ask another student volunteer to share how they solved the problem. Record what the student says, once again using an open number line.

Example of Student Thinking and Teacher Recording

Darren begins with four fingers and then has difficulty adding seven more. As the teacher, I suggested he use the counters to represent the number of seagulls in the problem.

I draw the following as Darren manipulates the counters:

Supporting Darren's thinking visually.

Then I ask, "What numbers should I write on the number line for this story?"

Darren responds, "Four and seven."

This transitions to the following diagram and I explain, "Darren started with four and then added seven more, ending up with eleven."

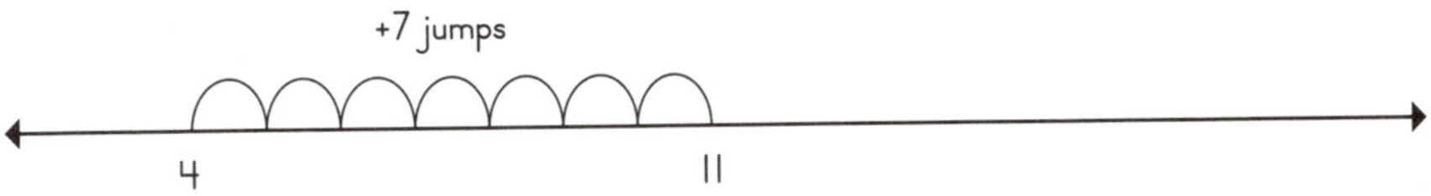

Recording Darren's thinking on an open number line.

10. Ask the student volunteer whether your recording (drawing) represents accurately what they said. If the student says it does not, ask the student to repeat their thinking as you change the recording accordingly.
11. Have one more student volunteer share their thinking. Record the student's thinking.

Example of Student Thinking and Teacher Recording

Monica added 6 + 1 jumps instead of seven individual jumps. So, as the teacher, I recorded the jumps according to the numbers Monica used. It could be that she was taking advantage of the *make-ten* strategy and just knew that four and six were ten, then added one more to finish adding seven.

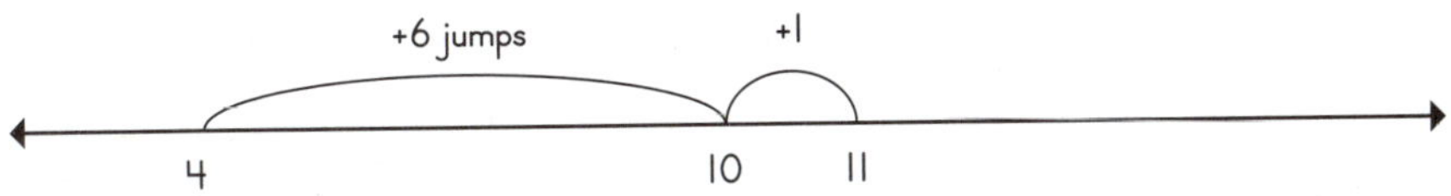

Recording Monica's thinking on an open number line.

Teaching Tip

Pacing

Begin with smaller numbers that can be modeled directly with counters. As students explain how to *add to* or *take from* a given set of objects, record their thinking on an open number line. After they are comfortable with using objects to show their solution processes, increase the size of the numbers in the story problem to multiples of ten to make the problems more suited to not having to count each object; use a visual diagram instead.

Teaching Tip

Things to Keep in Mind with Story Problems

1. Change the story problem as needed to reflect the interests, experiences, and names of students.
2. Present only one or two stories at a sitting.
3. Instead of telling students how to do the work ("you need to add"), ask them to explain the relationship of the numbers in the problem or use objects to represent the numbers.
4. Include additional information that may help to make the problem more realistic. It also may be helpful to include possibly nonessential information to make the problem either more engaging or to encourage children to evaluate the problem for information important to finding a solution.

Technology Tip

Using an Interactive Whiteboard

If an interactive whiteboard is available, display the problem first with blanks for the numbers. Then, use the pen tool to write in the missing numbers after the students have an understanding of what is happening in the problem.

12. Ask the student volunteer whether your recording (drawing) represents accurately what they said. If the student says it does not, ask the student to repeat their thinking as you change the recording accordingly.

13. To wrap up the problem, ask students to explain not only how many seagulls were sitting on the lunch table during recess, but what happened when there were so many seagulls out there. Students enjoy making suggestions for the ending of the story.

Examples of Student Thinking

Arturo: "One seagull grabbed some chips!"
Marcos: "They made a big noise."
Serena: "The kids screamed."

Problem Two: *Pencils*

14. On another day propose the second problem, *Pencils*. This problem can be presented to the whole class or a small group. Because the numbers are larger (chunks of tens), an open number line will be used to model and solve the problem instead of the number path.

15. Write the story problem where everyone in the class can see it. Include blanks instead of numbers, as shown here.

> Problem Two: Pencils
>
> The teacher has ____ pencils for the class. She sharpens ____ of them, but has to stop because the bell is ringing. She asks [insert name of student] to please sharpen the rest of them while the students enter the classroom. How many pencils does [insert name of student] sharpen for the class? Are there enough pencils for everyone?

16. Read the problem together with students.

17. Use the following questions to facilitate a productive discussion about the story problem:
    - "Tell the story problem in your own words."
    - "Share a different way to explain what is happening in the story."
    - "Tell your shoulder partner what this problem is all about."
18. Now insert numbers in the story problem as follows:

    Problem Two: Pencils

    The teacher has 40 pencils for the class. She sharpens 10 of them, but has to stop because the bell is ringing. She asks [insert name of student] to please sharpen the rest of them while the students enter the classroom. How many pencils does [insert name of student] sharpen for the class? Are there enough pencils for everyone?

19. Reread the story problem with students. Ask the following questions.
    - "Now that we know how many pencils are in the problem, what happened at the beginning?"
    - "What do we want to find out?"
    - "Who has a way to explain what to do?"
20. Have students solve the problem. Encourage students to act out parts of the problem so they can decide how many pencils the student will need to sharpen.
21. Ask a student volunteer to explain how they solved the problem. Record what the student says using an open number line.

Example of Student Thinking and Teacher Recording

Sarah: "She had forty pencils. Ten were sharp, but the rest were not sharp. I know that thirty and ten is forty, so thirty needed to be sharpened."

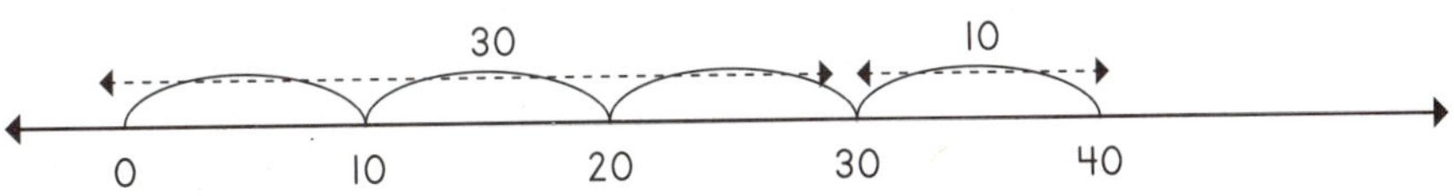

Recording Sarah's thinking on an open number line.

Teaching Tip

Why These Numbers?

The numbers in this story problem were chosen because they encourage students to make use of jumps of ten on an open number line when solving the problem.

Technology Tip

Using an Interactive Whiteboard

If an interactive whiteboard is available, use the line option to draw a straight line, then use the pen to label each end with arrows and numbers.

Teaching Tip

Extending Student Thinking

During the next few days, pass out one story problem a day from Story Problems: Solving Story Problems Involving Missing Numbers at the End (Reproducible 8). The numbers have been changed slightly from the ones used in this lesson. You may wish to change the context to match the interests of your students. If you do this, make sure you keep the structure the same. Have students work on the problems independently or in pairs.

Another option is to have students keep the problems in their math folder and work on them when they finish other math classwork. Later, after students are comfortable with using objects to show their solution process, you can increase the size of the numbers in the story problem to multiples of tens to make the problems more suited to not having to count each object, but using a visual diagram instead.

In this situation, I asked Sarah, "How did you know, so quickly, that thirty plus ten was forty?"

She responded, "I know three plus one is four, so thirty plus ten is forty."

22. Ask the student volunteer whether your recording (drawing) represents accurately what they said. If the student says it does not, ask the student to repeat their thinking as you change the recording accordingly.

23. Ask another student to share how they solved the problem. Record what the student says, once again using an open number line.

Example of Student Thinking and Teacher Recording

Jason: "You start with forty pencils and take ten over to the sharpener and sharpen them. Then you need to see how many you have that are not sharp."

40 – 10 = 30

Recording Jason's thinking on an open number line.

24. Ask the student volunteer whether your recording (drawing) represents accurately what they said. If the student says it does not, ask the student to repeat their thinking as you change the recording accordingly.

25. Now, ask students to explain not only how many pencils need to be sharpened, but how they know whether there are enough

pencils for the entire class. Students thus make sense of the problem as it relates to their own class.

26. After students have solved the problem, hold a whole-class discussion and encourage students to share their strategies as you record their thinking.

Math Talk

Teacher Moves to Support Student Thinking

The following is a list of tips to keep on hand when facilitating student thinking during problem solving.

- Make sure students understand the problem. *(Explain the problem in your own words.)*
- Encourage students to share with their peers what they do know about the problem. *(Tell the story problem to a shoulder partner in your own words.)*
- Promote reflection on the strategies that students use. *(Why does this method make sense? Would this work if _____?)*
- Explore what students already know. *(What do you know for sure?)*
- Encourage students to explore multiple strategies. *(How can you show this a different way? What can we do to act this out?)*
- Connect student's thinking to symbolic notation. *(What numbers should we write down? Where should we put them? Which one is more? Which one is less?)*
- Generate follow-up problems linked to the problem just solved. *(What if we had ____ instead?)*

Teacher Reflections

Making Sense of Story Problems

To help students make sense of story problems, I want to ensure they can explain the problem in their own words. This might mean asking three different students to explain the problem, so classmates can hear three different variations.

There might be language in the problem that is unfamiliar to some students. When students struggle to explain what is happening in the problem, it may be necessary to provide definitions and explanations of certain words. Typically, these are the words used in addition and subtraction story problems:

Addition	Subtraction
combined	minus
increased	less than
total of	less
sum	fewer than
added to	difference
together	decreased
plus	take away
	more than

All students need to be familiar with this kind of academic language to make sense of story problems. In some cases, words have a different or specific meaning in math that is very different from their use outside of math. For example, the word *sum* is often confused with the homonym *some*. For this reason, *total* is often used in primary classrooms when referring to *how many in all. Whole* and *hole* are another pair of homonyms that could be confusing. Other math-related multiple meaning (not homonyms) examples include *difference, face, mean,* and *plane.* When words such as these come up in the classroom, it is important to point out the distinctions. I like to have a math word wall that includes both the words and illustrations of the words; students can reference the word wall as needed.

Math Matters!

Supporting Students' Language Development

For more ideas in supporting students' language development in math class, see *Supporting English Language Learners in Math Class, Grades K–2* by Rusty Bresser, Kathy Melanese, and Christine Sphar (Math Solutions, 2009).

Instead of taking a word problem apart and looking for a few key words, I expand on the problem to make it more real and to clarify what is happening. I have found that I can help students understand a problem by asking them to pretend they are acting out the problem, using any available objects as props, and by inventing dialog to go with the actions.

Here is an example of questions I ask to prompt students when acting out a problem situation involving the confusing word *fewer:*

Luke has three fewer pencils than Jules. Jules has five pencils. How many pencils does Luke have?

- "Who wants to play the parts of Jules and Luke?"
- "What do you need to get for this problem?"
- "Jules, how many pencils are you supposed to have in this problem?"
- "Luke, do you have more than Jules or do you have fewer than Jules?
- "What do we mean by fewer?"
- "OK, so if you have fewer, does he have more?"
- "If they both have five pencils they would have the same number of pencils, but Luke has fewer."

When we listen carefully to how students explain a problem in their own words, we can determine how students are thinking about the meaning of the problem. In my early years as a teacher, I tried to make things clearer for students by eliminating many words in the problem and presenting just a few key words. However, I was the one who was doing the thinking! When students struggle to explain what a problem means, they are making sense of it. I am not doing it for them.

Thinking about Solution Strategies

I have seen three levels of solution strategies that students seem to use when solving problems such as those in Lesson 6, *Story Problems Involving Missing Numbers at the End*; Lesson 7, *Solving Story Problems Involving Missing Numbers in the Middle*; and Lesson 8, *Story Problems Involving Missing Numbers at the Beginning.*

Modeling with Objects

At first, students make use of counters or objects to stand for the things in the story problem. For example, when adding *4 + 5*, students count out a set of four objects and then count out a separate set of five objects, then they push the objects together and count all of them to obtain a total of nine.

Counting On or Counting Backwards

The second way students try to solve the same kind of addition problem is to use counting. For example, using the *4 + 5* problem, students begin with four and count on five more to get to nine. Some students might also begin at five and count on four more if they are starting to realize it is easier to count on from the larger number. Students often use their fingers to keep track of how many numbers they need to count on. This is an opportunity for the teacher to connect the counting-on strategy that children naturally do with their fingers to a number line recording, as in the previous examples.

Number Combinations

The third way students solve the same problem is to use *number combinations*. This method is usually developed after the other two methods are fully understood. To add *4 + 5*, for example, knowledge of the doubles might be used. ("I know that four and four are eight, so four and five would be nine." Or "I know that five and five are ten, so five and four would be nine because four is one less than five." We can see this connection in the example of Monica's thinking when Monica adds seven by adding six and then one more, because she is familiar with making ten.

I am convinced that the open number line is a useful bridge for students to use to represent their thinking. They can count one by one on the open number line if that method makes sense to them as a way to solve a problem. They can count on from a number and keep track of where they are. Finally, they can jump strategically on the open number line when they have internalized how friendly numbers such as tens can be used to make sense of a problem. The open number line is a tool for keeping track of computation while maintaining the size of the numbers, instead of looking only at ones and tens in an addition or subtraction algorithm.

Solving Story Problems Involving Missing Numbers in the Middle

Lesson 7

Overview

In this lesson, we explore a collection of problems that can be solved by students over a series of days. They are presented together so the reader can support students on a gradual journey to make sense of problem-solving situations that involve real things and simple number relationships. The open number line can be a tool to display and represent these story problem relationships.

The story problems in this lesson are referred to as *change-unknown* problems. What is a change-unknown problem? Let's consider three examples:

1. Two frogs are sitting on a log. Some more frogs hop up to join them. Now there are five frogs sitting on the log. How many frogs hopped up to join the first two? (Add to) 2 + ? = 5
2. Five pencils are in a cup on the table. I took some pencils. Now there are three pencils. How many pencils did I take? (Take from) 5 – ? = 3
3. Five marbles are in my pocket. Three are red marbles and the rest are green. How many marbles are green? (Put together/take apart, with one addend unknown) 3 + ? = 5 or 5 – 3 = ?

What do these three problems have in common? With change-unknown problems, the unknown number is the middle number. Students are given the whole and a known part

Time

15–20 minutes per problem, done over time as students become more comfortable dealing with making sense of problem situations

Materials

Story Problems: Solving Story Problems Involving Missing Numbers in the Middle (Reproducible 9), each problem on a separate sheet of paper, 1 set (4–6 problems) per student

counters and cubes (as needed to help students solve problems)

and required to conceptualize a missing part. Change-unknown problems are not commonly addressed in school. As a result, there is an increased need for using visual displays such as number lines to help students make sense of the problem and grapple with mathematical relationships such as addition and subtraction. The problems in this lesson are most appropriate for first and second graders.

Related Lessons

- L-5 Jumping by Ones, Fives, Tens, and Twenty
- L-6 Solving Story Problems Involving Missing Numbers at the End
- L-8 Solving Story Problems Involving Missing Numbers at the Beginning
- L-9 Solving Comparison Problems

Key Questions

- Who would like to explain what is happening in this story problem using your own words?
- What did we start with? What did we end up with?
- What happened?
- What do we want to know?

Teaching Directions

Problem One: *Pumpkins*

1. Write the following story problem where everyone in the class can see it. Include blanks instead of numbers, as shown here.

   Problem One: Pumpkins

   The teacher brings ____ pumpkins to school for the fall festival. Some children in the class bring in some

Teaching Tip

Why Blanks for Numbers?
When introducing story problems, it's recommended to use blanks for the numbers at first, so students do not try to figure out an answer right away. Rather, you want them to focus on understanding what's happening.

more pumpkins. Now there are ____ pumpkins in the classroom. How many pumpkins did the children bring to school?

What can you do with these ____ pumpkins for math time?

2. Read the problem together with students. This can be done with the entire class or in small groups. The paper with the problem and the counters can be made available to students after students have had an opportunity to make sense of the problem.

3. Use the following statements to facilitate a productive discussion about the story problem:

   - "Tell the story problem in your own words."
   - "Share a different way to explain what is happening in the story."
   - "Tell your shoulder partner what this problem is all about."

4. Now insert numbers in the story problem as follows:

   Problem One: Pumpkins

   The teacher brings 2 pumpkins to school for the fall festival. Some children in the class bring in some more pumpkins. Now there are 5 pumpkins in the classroom. How many pumpkins did the children bring to school?

   What can you do with these 5 pumpkins for math time?

5. Reread the story problem with students. Ask the following questions:

   - "Now that we know how many pumpkins are in the problem, what happened in the middle?"
   - "What do we want to find out?"
   - "Who has a way to explain what to do?"

Technology Tip

Using an Interactive Whiteboard

If an interactive whiteboard is available, display the problem first with blanks for the numbers. Then, use the pen tool to write in the missing numbers after students have an understanding of what is happening in the problem.

Teaching Tip

Why These Numbers?

Smaller numbers have been chosen for these story problems during the whole-class introduction so that students can model what's happening more easily using representational objects, which helps them determine the missing number. Later, larger number sums can be used, after students understand part-part-whole relationships.

Teaching Tip

Making Sense of Story Problems

To help students make sense of story problems, it's critical to spend time ensuring they understand what is happening in the problem. Consider this example; the following story problem was presented to a first grader:

Twelve mice live in a house. Nine live upstairs. How many live downstairs?

It soon became apparent that the first grader was grappling with the meaning of the problem. She thought there were twelve mice and nine more. She counted out twelve counters and then counted out another nine counters to represent the mice, yet she didn't know what to do with this information. At this point, the teacher realized the first grader might not be understanding the separation of upstairs and downstairs. He changed the story to include a physical action that suggested separation:

Twelve mice live in a house. Nine of those mice are going to go upstairs to watch TV. How many mice are still downstairs?

Now the student was able to separate the nine that were going upstairs from the other twelve mice, leaving three mice (Jacobs & Ambrose 2008).

6. Have students solve the problem. Suggest they act out the problem. You can ask, "What can we use for the pumpkins? Who can play the role of the teacher? How many pumpkins will the teacher have?" The teacher can prompt with questions about the story situation. Consider, "Are they bringing more pumpkins?"
7. Ask a student volunteer to explain how they solved the problem. Record what the student says using drawings as well as an open number line. In this case, you are the one who inserts the symbol (+) into the representation when the student talks about the two pumpkins and the other pumpkins. You make the connection to the visual representation while the student explains the problem.

Example of Student Thinking and Teacher Recording

Chelsea: "We started with two pumpkins and then there were three more to be five pumpkins."

As the teacher, I drew a representation of the pumpkins:

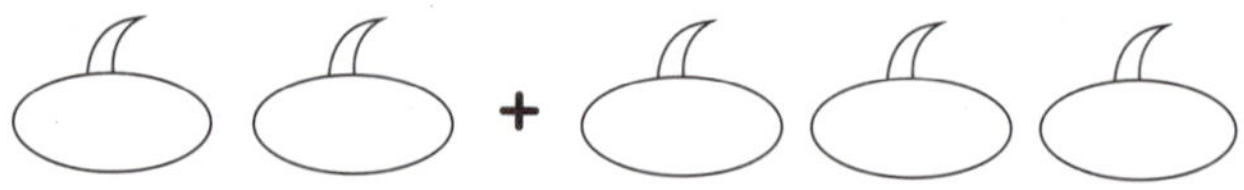

Recording Chelsea's thinking as a drawing.

I then checked with Chelsea to make sure my drawing represented her thinking accurately. "So, is this what you're saying, Chelsea?"

Chelsea confirmed, "I started with the two pumpkins and then put three more in until I got to five."

I then proceeded to record Chelsea's thinking on an open number line, using the plus sign for the words *and more* in the description.

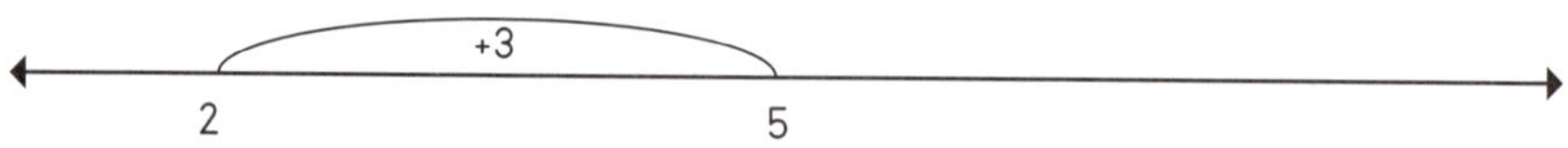

Recording Chelsea's thinking on an open number line.

8. Ask the student volunteer whether your recording (drawing) represents accurately what they said. If the student says it does not, ask the student to repeat their thinking as you change the recording accordingly.

9. Ask another student volunteer to share how they solved the problem. Record what the student says, once again using an open number line.

Example of Student Thinking and Teacher Recording

Rex: "We had five pumpkins after some more were brought to school, but we started with two, so we take three away to get back to the two we had at the beginning."

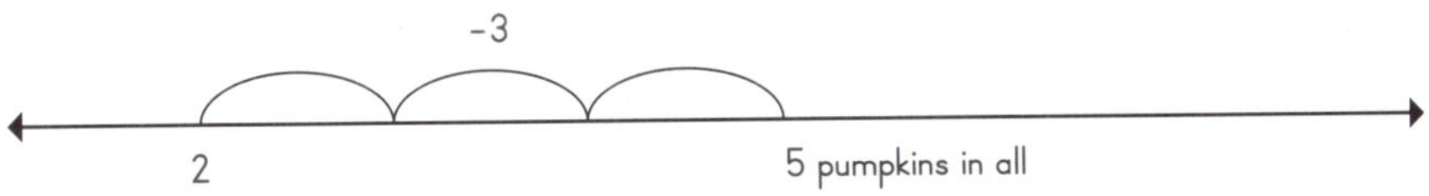

Recording Rex's thinking on an open number line.

10. Ask the student volunteer whether your recording (drawing) represents accurately what they said. If the student says it does not, ask the student to repeat their thinking as you change the recording accordingly.

11. Ask if someone else has another way to tell the pumpkin story using counters. Record that student's thinking.

Example of Student Thinking and Teacher Recording

Brendan: "I started with two counters and added one and one and one until I got to five."

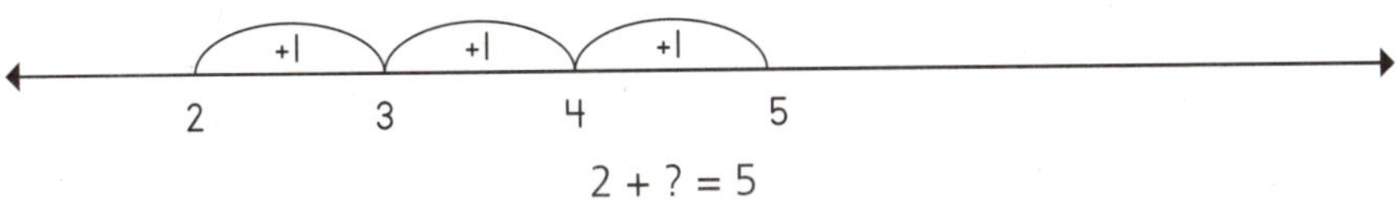

Recording Brendan's thinking on an open number line.

Teacher: Note that the question mark in the number sentence is to show that "at first you

A Child's Mind . . .

Subtracting Jumps

Contrary to Chelsea's strategy, if a student subtracts three jumps from five to solve the problem, record the jumps using the subtraction sign.

Math Matters!

The Equal Sign

When thinking about the total at the beginning of a situation, the matching equation looks like this: $5 = 3 + 2$. These kinds of equations help children understand that the equal sign does not always mean *makes* or *results* in. It can also mean *is the same number as*. Later, students need this understanding of the meaning of the equal sign to make sense of algebraic expressions such as $2 + 3 = 4 + 1$.

Math Matters!

Using Bar Models

The concept of bar models originally came from Singapore math textbooks and was designed to help students visualize problems by turning abstract words into easy-to-understand pictorial models. If students look only at key words and see the word *more*, they may be confused about what the problem is asking them to do. They may think the problem means to add. By drawing a representation model of the problem, they can understand the problem more clearly.

Here's an additional example:

There are 7 boys and 21 girls in a class. How many more girls than boys are there?

To illustrate and understand the problem, draw two bars. There are more girls than boys, so draw a shorter bar to represent the boys and a longer bar to represent the girls. The question mark indicates what we are trying to find out—the difference between the number of boys from the number of girls. Many students are more comfortable with visual representations of abstract word relationships.

21 girls	
7 boys	?

didn't know how many to add on until you got to five."

12. Now share with students another way to think about solving story problems: using a bar model. Draw the following bar model and explain to students that this shows the whole pumpkin story.

whole	
part	part

Point to the top bar and say, "This bar is the whole group of pumpkins." Point to the bottom bars and say, "And these two cells are the parts that, when they go together, make up the whole, or total, amount."

"So, if we put numbers on the bars, it would look like the following."

5 pumpkins	
2 pumpkins	Some more pumpkins

13. To wrap up the problem, ask, "What could we do with the pumpkins during math time?" Students enjoy making suggestions for how to measure, weigh, and rearrange the five pumpkins.

Problem Two: *Sticks of Chalk*

14. On another day, propose a second problem: *Sticks of Chalk*. Write the story problem where everyone in the class can see it. Include blanks instead of numbers, as shown here.

Problem Two: Sticks of Chalk

We have __ sticks of colored chalk for the chalk art festival on the blacktop outside our classroom. The principal brings more sticks of colored chalk for us to use. Now we have __ sticks of chalk. How many sticks of chalk did the principal bring?

What should we draw on the blacktop for the chalk art festival?

15. Read the problem together with students. This can be done in whole or small groups.

16. Use the following statements to facilitate a productive discussion about the story problem:

   - "Tell the story problem in your own words."
   - "Share a different way to explain what is happening in the story."
   - "Tell your shoulder partner what this problem is all about."

17. Now insert numbers in the story problem as follows:

   Problem Two: Sticks of Chalk

   We have 30 sticks of colored chalk for the chalk art festival on the blacktop outside our classroom. The principal brings more sticks of colored chalk for us to use. Now we have 52 sticks of chalk. How many sticks of chalk did the principal bring?

   What should we draw on the blacktop for the chalk art festival?

18. Reread the story problem with students. Ask the following questions:

   - "Now that we know how many sticks of chalk are in the problem, what happened in the middle?"
   - "What do we want to find out?"
   - "Who has a way to explain what to do?"

19. Have students solve the problem. To support student understanding, the teacher can provide three baggies of ten crayons that can represent the chalk in the beginning of the problem. Children could use ten connecting cubes to represent ten sticks of chalk. Some children might draw ten tally marks to show ten sticks of chalk.

20. Ask a student volunteer to explain how he or she solved the problem. Record what the student says using an open number line.

Teaching Tip

Things to Keep in Mind with Story Problems

1. Change the story problem as needed to reflect the interests, experiences, and names of students.
2. Present only one or two stories at a sitting.
3. Instead of telling students how to do the work ("you need to add"), ask them to explain the relationship of the numbers in the problem or use objects to represent the numbers.
4. Include additional information that may help to make the problem more realistic. It also may be helpful to include possibly nonessential information to make the problem either more engaging or to encourage children to evaluate the problem for information important to finding a solution.

Math Matters!

Adding and Subtracting Two-Digit Numbers

Larger numbers in story problems can be thought of as units of ten or tens plus ones. To become fluent with adding and subtracting two-digit numbers, students need to be confident in adding multiples of ten to any two-digit number. The time they spend practicing this skill pays off in increased understanding and fluency with addition and subtraction of multiple-digit numbers.

Technology Tip

Using an Interactive Whiteboard

If an interactive whiteboard is available, use the line option to draw a straight line, then use the pen to label each end with arrows and numbers.

Example of Student Thinking and Teacher Recording

Tomas: "We had thirty sticks of chalk at first, but then we ended up with fifty-two, so we need to find out how many the principal gave us. If he gave us ten, we would have forty. If he gave us twenty, we would have fifty, so we know he gave us more than twenty."

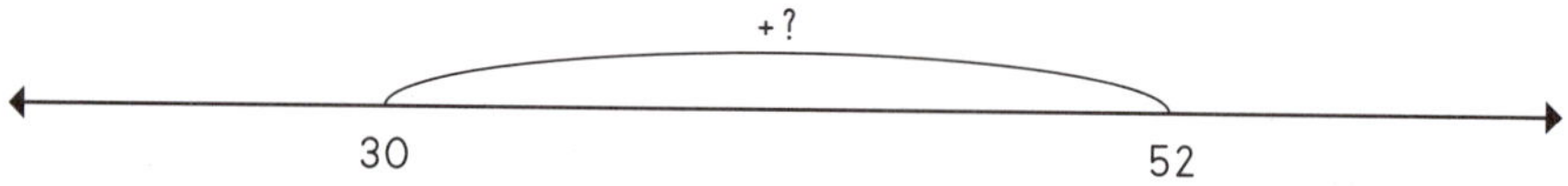

30 + ? = 52

Recording Tomas' thinking on an open number line.

A Child's Mind . . .

Hidden Inside

In decomposing/composing story number problems like these, students shift from thinking about the total to thinking of the addends. Working with different numbers helps students learn *number families*, which they can use when adding and subtracting. Eventually, with more experience, students are able to think of numbers that are "hidden inside" the total number while thinking about the number at the same time.

21. Ask the student volunteer whether your recording (drawing) represents accurately what he or she said. If the student says it does not, ask the student to repeat his or her thinking as you change the recording accordingly.

22. Ask other students to explain how they could jump from 30 to 52. Some may jump by twenty and then two more. Others may make two jumps of ten and then two ones to get to 52. Others may need to jump by ones all the way to 52.

Examples of Student Thinking and Teacher Recording

Olivia: "We jumped by twenty and then two more."

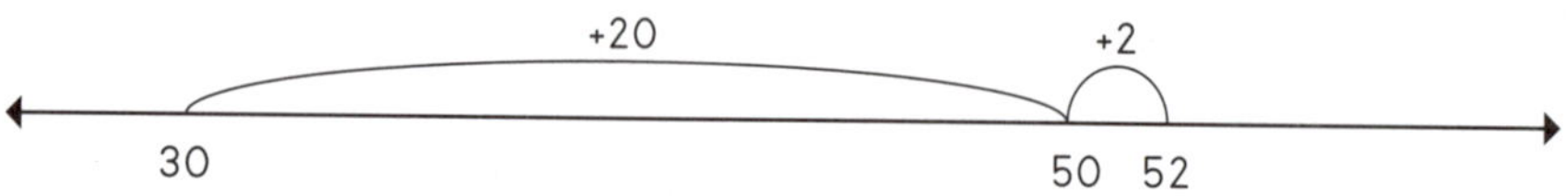

Recording Olivia's thinking on an open number line (*30 + 20 = 50* and *50 + 2 = 52*).

Vinod: "We made two jumps of ten and then two ones to get to fifty-two."

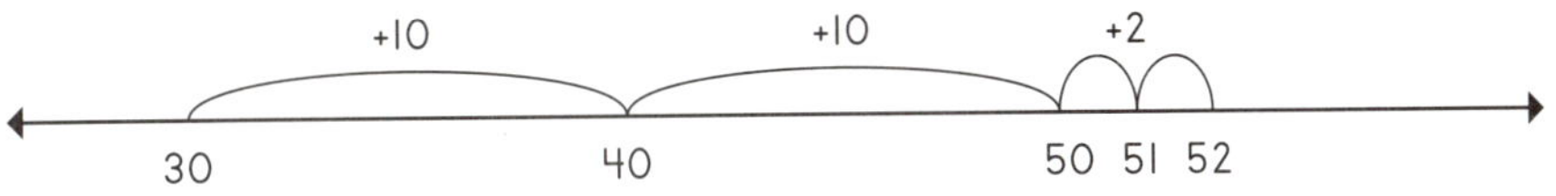

$30 + (10 + 10 + 2) = 52$

Recording Vinod's thinking on an open number line ($30 + (10 + 10) + (1 + 1) = 52$).

23. Now draw the following bar model and explain to students that this shows the relationship of the numbers. Say, "This is another way to illustrate this story problem. The bar at the top stands for all the chalk. The first part below the whole bar is the chalk we already had—thirty—and the second part is the chalk the principal brought. Put the two bars together and you have all the chalk: fifty-two sticks."

whole	
part	part

52 sticks of chalk	
30 sticks of chalk	Some more sticks of chalk

Math Matters!

Recording Equations

The parentheses in the recording representing Vinod's thinking are inserted by the teacher, but they do not need to be addressed at this time. Students become familiar with this notation and what it means in this context without having to use it themselves. More exposure over time helps them make sense of the abstract symbols. For younger children, this equation can also be expressed as $30 + 20 + 2 = 52$ or $30 + 10 + 10 + 2 = 52$.

24. Wrap up the problem by asking, "What should we draw on the blacktop for the art festival?" The final question in this problem is to bring meaning to the story. Students will likely have many suggestions for what to draw.

Problem Three: *The Front Row*

25. On another day you can introduce another story problem with missing information in the middle. Write this story problem where everyone in the class can see it. Include blanks instead of numbers, as shown here.

Problem Three: The Front Row

The teacher wants to have ____ students sit in the front row. There are ____ boys in the front row. How many girls should sit in the front row with these boys? Draw a picture of the front row and use B for boy and G for girl. How would you arrange the boys and girls? Why?

This is a picture frame for the front row:

26. Read the problem together with students. Say, "Let's pretend this is the first row of seats."

27. Use the following statements to facilitate a productive discussion about the story problem:

    - "Tell the story problem in your own words."
    - "Share a different way to explain what is happening in the story."
    - "Tell your shoulder partner what this problem is all about."

28. Now insert numbers in the story problem as follows:

    Problem Three: The Front Row

    The teacher wants to have 5 students sit in the front row. There are 3 boys in the front row. How many girls should sit in the front row with these boys? Draw a picture of the front row and use B for boy and G for girl. How would you arrange the boys and girls? Why?

    This is a picture frame for the front row:

29. Reread the story problem with students. Ask the following questions.

    - "What happened at the beginning?"
    - "What do we want to find out?"
    - "Who has a way to explain what to do?"

Math Matters!

Several Ways

In the *Front Row* Problem, there are several ways for students to think about how three boys and two girls can be arranged in one row. Although there is an answer for the number of girls (two), the second part of the problem involves thinking about different arrangements for this particular combination. The "equivalent expression" of $3 + 2 = 5$ makes this problem into an opportunity to discuss and model a variety of ways to show different number combinations. This kind of activity helps students make sense of what $3 + 2$ might look like with boys and girls as parts of the whole.

Examples of Student Thinking

Deanna: "The teacher wants there to be five kids in the row."

Mario: "There are three boys in the front row."

Teacher: "What do we want to find out?"

Stephanie: "How many girls are going to be in the front row?"

Teacher: "Who has a way to explain what to do?"

Rex: "Start filling in the row with three boys and then count on girls until you get to five."

30. Have students solve the problem. Say, "Talk with your shoulder partner about how you think the three boys and the rest of the girls should be seated. You can sketch out the problem by drawing on scratch paper."
31. Ask a student volunteer to explain how they thought about the number of boys and girls in the front row. Record what the student says using an open number line.

Example of Student Thinking and Teacher Recording

Mona: "I went to three and then said 'four, five' and got two more."

Recording Mona's thinking on an open number line ($3 + 2 = 5$).

32. Ask the student volunteer whether your recording (drawing) represents accurately what he or she said. If the student says it does not, ask the student to repeat his or her thinking as you change the recording accordingly.
33. Ask another student to share how they know that two more girls need to join the boys in the front row. Ask the student to show his or her thinking using counters (or other students!).
34. Now draw a bar model and explain that the top bar represents all the students in the front row, the first part is the group of boys, and the second part is the group of girls. When they are put together, they make the whole group.

whole (all of the children in the row)	
part (boys)	part (girls)

Now say, "Here is a diagram that portrays the key elements of the story problem."

5 children in the front row	
3 boys	2 girls

35. Support students in extending their thinking. Ask, "How could the teacher arrange the students in the row? For example, might she put the boys next to each other?"
36. Use cubes of different colors (one color to represent the girls and another to represent the boys) to act out this problem.
37. After students have solved the problems, hold a whole-class discussion and encourage students to share their strategies as you record their thinking and support students in finding ways to show their thinking on the paper sheet.

Math Matters!

Exploring Different Arrangements
Finding ways to explore different arrangements strengthens children's understanding of the meaning of the sum (five, in this example). Permutations is a topic children study later in mathematics.

Teaching Tip

Extending Student Thinking
During the next few days, distribute one story problem a day from Story Problems: Solving Story Problems Involving Missing Numbers in the Middle (Reproducible 9). The numbers have been changed slightly from the ones used in this lesson. You may wish to change the context to match the interests of your students. If you do this, make sure you keep the structure the same. Have students work on the problems independently or in pairs.

Another option is to have students keep the problems in their math folder and work on them when they finish other math classwork. Later, after students are comfortable with using objects to show their solution process, you can increase the size of the numbers in the story problem to multiples of tens to make the problems more suited to not having to count each object, but using a visual diagram instead.

Teacher Reflections

The Front Row Problem

When I tried the *Front Row* Problem with first graders, they were not at all systematic about making arrangements of boys and girls. Some wanted the girls to be able to sit next to each other, whereas others tried to make an alternating pattern so the boys would not sit next to each other. I decided to have students use red cubes to represent the girls and blue to represent the boys. I asked students to lay out each new arrangement on their 1-by-5 frame so that it had two girls and three boys. We colored in the frames and tried to find different arrangements using the cubes first, then comparing the colored papers to determine whether there were duplicates. Here are some ways the children can be arranged:

BBBGG	BGGBB
BBGBG	GBGBB
BBGGB	GGBBB
BGBBG	GBBGB
BGBGB	GBBBG

All of this work strengthened students' understanding of the idea that $3 + 2 = 5$, because they dealt with these combinations over and over again. I think some students were pretty amazed there could be so many different ways to arrange one row of children.

Building an Awareness of Problem Types

Learning about different types of problems helped me become more aware of the fact that, often in school, children are presented with a narrow spectrum of problem situations that only involve joining or separating with the total unknown. During my early teaching years, I typically used addition and subtraction problems in which the unknown number was at the end of the problem. When I learned of a wider variety of addition and subtraction problems, I realized how important it is for children to grapple with different types of problems. How could I expect children to make sense of a problem when they have never seen one posed with the unknown numbers in the middle or in the beginning?

I've also learned that when children solve problems, sometimes in more than one way, and report how the problems were solved to their classmates and to me, it's an opportunity for us to listen and question until we also understand the solutions.

Lesson 8

Solving Story Problems Involving Missing Numbers at the Beginning

Time

15–20 minutes per problem, done over time as students become more comfortable dealing with making sense of problem situations

Materials

Story Problems: Solving Story Problems Involving Missing Numbers at the Beginning (Reproducible 10), each problem on a separate sheet of paper, 1 set (6 problems) per student

counters

interlocking cubes, 30 (15 red and 15 blue), per group

index cards, 10 per group (as needed to help students solve problems)

Overview

In this group of story problem lessons we explore a collection of problems that can be solved by students over a series of days. They are presented together so the reader can support students on a gradual journey to make sense of problem-solving situations that involve real things and simple number relationships. The open number line can be a tool to display and represent these story problem relationships.

The group of story problems in this lesson are referred to here as *start-unknown* problems. What is a start-unknown problem? Let's consider three examples:

1. Some frogs are sitting on a log. Three more frogs hop up to join them. Now there are 5 frogs sitting on a log. How many frogs were on the log before the others joined them? (Add to) ? + 3 = 5
2. Some pencils are in a cup on the table. I took 2 pencils. Now there are 3 pencils. How many pencils were in the cup before I took 2 of them? (Take from) ? − 2 = 3
3. I have 5 marbles. How many can I put in my front pocket and how many can I put in my back pocket? (Put together/take apart, with both addends unknown)

What do these three problems have in common? In start-unknown problems, the unknown number is at the beginning. The structure of these problems involves an unknown part (or two parts), a known part, and a total that is known. There is an increased need for using visual displays such as number lines to help students make sense of these not-commonly-taught problems and grapple with the relationships between addition and subtraction. This lesson is most appropriate for first and second graders.

Related Lessons

- L-6 Solving Story Problems Involving Missing Numbers at the End
- L-7 Solving Story Problems Involving Missing Number in the Middle
- L-9 Solving Story Comparison Problems

Key Questions

- Who would like to explain what is happening in this story problem using your own words?
- What did we start with? What did we end up with?
- What happened?
- What do we want to know?

Teaching Directions

Problem One: *Building with Blocks*

1. Write the following story problem where everyone in the class can see it. Include blanks instead of numbers, as shown here.

   Problem One: Building with Blocks

   Some students in the classroom are making block buildings. ____ more students join them on the floor to help. Now there are ____ students making block buildings. How many students were making block buildings on the floor before the other students joined them?

   What kinds of buildings do you think they are making?

2. Read the problem together with students.

3. Use the following statements to facilitate a productive discussion about the story problem:
   - "Tell the story problem in your own words."
   - "Share a different way to explain what is happening in the story."
   - "Tell your shoulder partner what this story problem is all about."
4. Now insert numbers in the story problem as follows:

   Problem One: Building with Blocks

   Some students in the classroom are making block buildings. Three more students join them on the floor to help. Now there are 5 students making block buildings. How many students were making block buildings on the floor before the other students joined them?

   What kinds of buildings do you think they are making?
5. Reread the story problem with students. Ask the following questions:
   - "Now that we know how many students are in the problem, what happened at the beginning?"
   - "What do we want to find out?"
   - "How can we explain what to do?"
6. Have students solve the problem. Suggest they act out the problem. What can students use for the blocks in the story? You can prompt with questions about the story situation, such as, How many children are there now? What happened in the middle?
7. Ask a student volunteer to explain how they solved the problem. Record what the student says using an open number line. In this case, you are the one who inserts the symbol (+) into the representation.

Teaching Tip

Why These Numbers?
Smaller numbers were chosen for this story problem during the whole-class introduction so that students can model what's happening more easily using representational objects, which help them determine the missing number. Later, larger numbers can be used, after students understand part-part-whole relationships.

Teaching Tip

Recording Using a Mathematical Symbol
In this case, the teacher is the one who inserts the symbol (+) into the representation when the student talks about *three more* students coming to the table. You make the connection to the visual representation while the student explains the problem.

Example of Student Thinking and Teacher Recording

Deanna: "We don't know how many were playing in the beginning. We know that three came."

Rex: "And, we know there's five playing at the end."

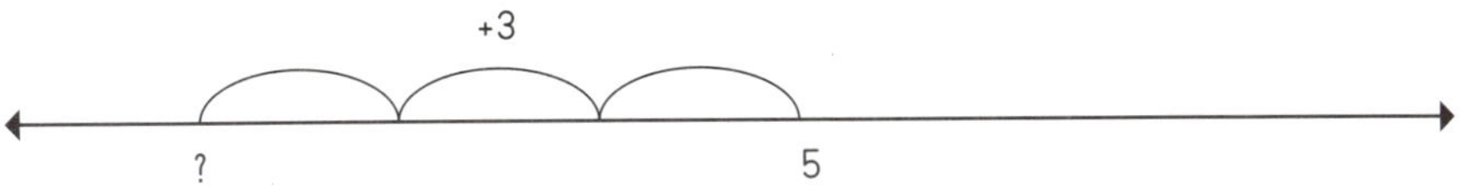

? + 3 = 5

Recording Deanna and Rex's thinking on an open number line.

8. Ask the student volunteer whether your recording (drawing) represents accurately what they said. If the student says it does not, ask the student to repeat their thinking as you change the recording accordingly.
9. Ask another student volunteer to share how they solved the problem. Record what the student says, once again using an open number line.

A Child's Mind . . .

Subtracting Jumps

Contrary to Deanna and Rex's strategy, if a student subtracts three jumps from five to solve the problem, record the jumps using the subtraction sign.

Example of Student Thinking and Teacher Recording

Maggie: "I know that there were five there at the end. Three came, so there were two there at the beginning because five take away three is two."

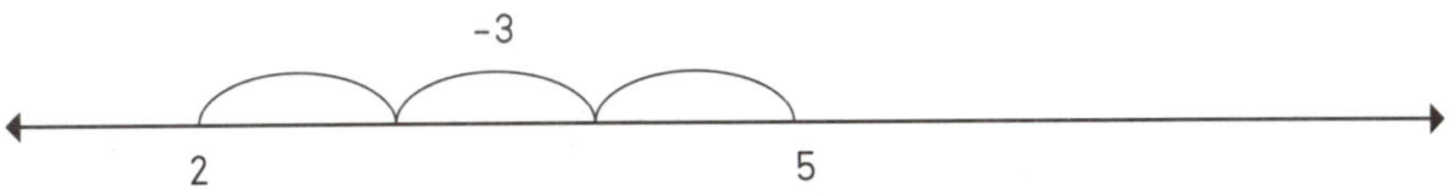

Recording Maggie's thinking on an open number line.

Math Matters!

Fact Families

Because the beginning number is unknown, the relationship of addition to subtraction is important to make sense of this kind of story problem. We know that "some" children and three more will total five. All of these equations relate to one another:

$5 - 3 = 2$, $3 + 2 = 5$, and $5 = 3 + 2$.

10. Ask the student volunteer whether your recording (drawing) represents accurately what they said. If the student says it does not, ask the student to repeat their thinking as you change the recording accordingly.

11. Now ask, "How can we think about the number of students who were there at first? Could it have been one student? Four students? Eight students?" Suggest that students use counters (or other students!) to act out the situation.

12. Now share with students another way to think about solving problems: using a bar model. Draw the following bar model and explain to students that it shows the whole story.

whole	
part	part

Say, "So, here is a diagram, called a bar model, that shows what we know in the story problem. The top bar represents the whole—the total number of children. The bottom bars are the parts that, when they go together, make up the whole."

13. Now change "whole," "part," and "part" to the actual story:

5 students	
some children	3 more students

Say, "The whole group in this problem is the students who are making buildings with blocks. We know there are five students at the end. We also know, from the second part of the problem, that three students came to help the first group. But, we don't know how many there were in the first part."

14. Now ask, "How do we know how many students are there at the beginning?"

Examples of Student Thinking

Mario: "You'll have five at the end if you start with two and three more come to play."

Olivia: "Because two and three is five."

15. To wrap up the problem, ask, "What types of buildings do you think the students are making?" Your students will enjoy making suggestions to describe the block buildings!

Problem Two: *Juice Boxes*

16. On another day, propose the second problem, *Juice Boxes*. Write the story problem where everyone in the class can see it. Include blanks instead of numbers, as shown here.

    Problem Two: Juice Boxes

    Some juice boxes are on a table in our classroom. The snack helper takes __ juice boxes to the lunch area. Now there are __ juice boxes on the table. How many juice boxes were on the table before? Are there enough juice boxes for our whole class?

17. Read the problem together with students. This can be done with the entire class or in small groups. The paper with the problem and the counters can be made available to students after they have had an opportunity to make sense of the problem.

18. Use the following statements to facilitate a productive discussion about the story problem:

    - "Tell the story problem in your own words."
    - "Share a different way to explain what is happening in the story."
    - "Tell your shoulder partner what story this problem is all about."

> **Teaching Tip**
>
> **Why Blanks for Numbers?**
> When introducing story problems, it's recommended to use blanks for the numbers at first, so students do not try to figure out an answer right away. Rather, you want them to focus on understanding what's happening.

Technology Tip

Using an Interactive Whiteboard

If an interactive whiteboard is available, display the problem first with blanks for the numbers. Then, use the pen tool to write in the missing numbers after students have an understanding of what is happening in the problem.

Teaching Tip

Things to Keep in Mind with Story Problems

1. Change the story problem as needed to reflect the interests, experiences, and names of students.
2. Present only one or two stories at a sitting.
3. Instead of telling students how to do the work ("you need to add"), ask them to explain the relationship of the numbers in the problem or use objects to represent the numbers.
4. Include additional information that may help to make the problem more realistic. It also may be helpful to include possibly nonessential information to make the problem more engaging or to encourage children to evaluate the problem for information important to finding a solution.

19. Now insert numbers in the story problem as follows:

    Problem Two: Juice Boxes

    Some juice boxes are on a table in our classroom. The snack helper takes 10 juice boxes to the lunch area. Now there are 17 juice boxes on the table. How many juice boxes were on the table before?

    Are there enough juice boxes for our whole class?

20. Reread the story problem with students. Ask the following questions:
    - "Now that we know how many juice boxes are in the problem, what happened at the beginning?"
    - "What do we want to find out?"
    - "Who has a way to explain what to do?"
21. Have students solve the problem. You may need to change the numbers in the problem if you have more than 27 students in your class. Even in a make-believe story problem, the students all want to have a juice box for snack.
22. Ask a student volunteer to explain how they solved the problem. Record what the student says using an open number line.

Example of Student Thinking and Teacher Recording

Olivia: "I know that there's more because they took ten outside and there were still seventeen on the table. Maybe she couldn't carry more than ten at a time."

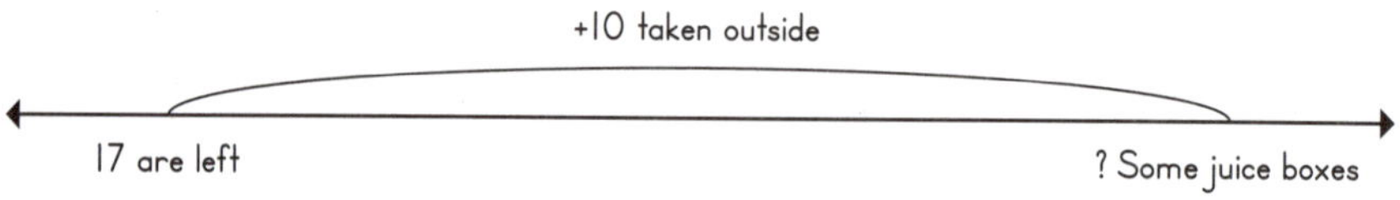

Recording Olivia's thinking on an open number line.

23. Ask the student volunteer whether your recording (drawing) represents accurately what they said. If the student says it does not, ask the student to repeat their thinking as you change the recording accordingly.

24. Now ask students to explain how they know there are more juice boxes at the beginning of the story. Encourage students to use counters to show what is happening in the story.

25. Refer to the bar model again. Explain how the bar model shows the relationship of juice boxes from the story. Say, "The whole part is that there are some juice boxes at the beginning of the story. The first part is that the snack helper took ten boxes to the lunch area. The second part is that seventeen boxes are left on the table."

whole	
part	part

Some juice boxes	
10 juice boxes taken to lunch area	17 juice boxes on the table

26. Ask, "So how many juice boxes are there at the beginning? How do we know?"

27. Wrap up the problem by asking, "How many juice boxes do we need for our class? Is twenty-seven enough?"

Problem Three: *Table Seating*

28. On another day, you can continue with the third problem, *Table Seating*. You will need the following materials for this problem: thirty interlocking red and blue cubes and ten index cards. Write the story

Math Talk

Teacher Moves to Support Student Thinking

The following is a list of tips to keep on hand when facilitating student thinking during problem solving.

- Make sure students understand the problem. *(Explain the problem in your own words.)*
- Encourage students to share with their peers what they do know about the problem. *(Tell the story problem to a shoulder partner in your own words.)*
- Promote reflection on the strategies that students use. *(Why does this method make sense? Would this work if _____?)*
- Explore what students already know. *(What do you know for sure?)*
- Encourage students to explore multiple strategies. *(How can you show this a different way? What can we do to act this out?)*
- Connect student's thinking to symbolic notation. *(What numbers should we write down? Where should we put them? Which one is more? Which one is less?)*
- Generate follow-up problems linked to the problem just solved. *(What if we had ____ instead?)*

problem where everyone in the class can see it. Include blanks instead of numbers, as shown here.

Problem Three: Table Seating

The teacher wants to have ___ students sitting at each table. How many boys and how many girls can he put at one table? Make a list for the teacher, using B for boy and G for girl.

Why did you arrange the students the way you did?

29. Read the problem together with students when they are seated at their table in groups.
30. Use the following statements to facilitate a productive discussion about the story problem:
    - "Tell the story problem in your own words."
    - "Share a different way to explain what is happening in the story."
    - "Tell your shoulder partner what this problem is all about."
31. Now insert numbers in the story problem as follows:

Problem Three: Table Seating

The teacher wants to have 5 students sitting at each table. How many boys and how many girls can he put at one table? Make a list for the teacher, using B for boy and G for girl.

Why did you arrange the students the way you did?

32. Reread the story problem with students. Ask the following questions:
    - "Now that we know how many students are at each table, what do we do?"
    - "What do we want to find out?"
    - "What are some of the ways to arrange the students?"

Math Matters!

Multiple Correct Answers

This problem is different from all the others because multiple correct answers are possible. It still fits into the category of a part-part-whole relationship, but the use of the number line here is limited to different ways we can arrange five items that are lined up on the number line in the first five places.

Teaching Tip

How Many Students Do I Call On?

Call on enough students to gain a variety of answers. From these answers, students should understand that multiple answers are possible for this problem. This exchange leads to conversation about what those many answers could be.

Examples of Student Thinking

Brendan: "He wants to have five kids at a table. So, there will be five and five and five and five and five and five at the tables."

Tomas: "He might have five boys at one table."

Olivia: "And he might have five girls at one table."

Maggie: "He could have boys and girls at some tables."

33. Share with students that you want to help them think of the different ways—not just all boys or all girls—but different ways to make up five students at a table. Pass out thirty interlocking red and blue cubes and ten index cards to each group of students, and explain, "The cards are the tables, the red cubes are girls, and the blue cubes are boys. Use these items to figure out different student arrangements."
34. Give students time to work out the different arrangements in their small groups. Circulate among the students and observe and note those students you might ask to share their arrangements to the whole class.
35. Bring the students back together as a whole class and ask, "Who has a way to explain what the tables could look like?"

Examples of Student Thinking

Rex: "We could have five boys at a table."

Olivia: "Or we could have five girls."

Tomas: "It's too hard, because we could have some girls and some boys."

Teacher: "Is there some way you could show your list of five boys using the letter B?

Rex: "You could write BBBBB."

Teacher: "How then would you show the five girls?"

Olivia: "Write GGGGG."

Math Talk

Small Groups
Working in small groups allows students to talk through their thinking and hear from their classmates. This collaboration helps them share ideas and, like think-pair-share, allows students to rehearse what they want to say before voicing it in front of the whole class.

Teaching Tip

Extending Student Thinking

On another day, you can distribute individual or several story problems from Story Problems: Solving Story Problems Involving Missing Numbers in the Beginning (Reproducible 10). The numbers have been changed slightly from the ones used in this lesson, and there are more story problems in the reproducible than described in this lesson. You may wish to change the context to match the interests of your students. If you do this, make sure you keep the structure the same. Have students work on the problems independently or in pairs.

Another option is to have students keep the problems in their math folder and work on them when they finish other math classwork. Later, after students are comfortable with using objects to show their solution processes, you can increase the size of the numbers in the story problem to multiples of tens to make the problems more suited to not having to count each object, but using a visual diagram instead.

Teacher: "What if you had one boy in the group?"

Tomas: "Then you'd have BGGGG."

36. To wrap up, ask students to record their thinking. Say, "Now see if you can figure out other ways to arrange the students."

Example of Student Thinking

When asked to record his thinking, one student comes up with this drawing for $5 = 3 + 2$:

37. After students have solved the problems, hold a whole-class discussion and encourage students to share their strategies as you record their thinking.

Teacher Reflection

Problems with More Than One Correct Answer

I presented the *Table Seating* Problem to a first-grade class in March:

> The teacher wants to have _5_ students sitting at each table. How many boys and how many girls could he put at one table? Make a list for the teacher, using B for boy and G for girl.
>
> Why did you arrange the students the way you did?

As mentioned earlier, students shared the following information with me:

> Brendan: "He wants to have five kids at a table. So, there will be five and five and five and five and five at the tables."
>
> Tomas: "He might have five boys at one table."
>
> Olivia: "And he might have five girls at one table."
>
> Maggie: "He could have boys and girls at some tables."

After a few comments, the students stalled. They just didn't know what to do with the information. Because the problem has multiple solutions, I wanted to help them find a way to list all the possible solutions.

I decided to give each group of students a collection of blue and red interlocking cubes. I explained that the blue cubes represented the boys and the red cubes stood for the girls. I also gave each group index cards and told them that these could represent the tables. They were to make up table groups with five cube "kids" at each "table." I showed them one "table" with five blue cubes and told them this was one of the arrangements they had mentioned earlier. I didn't think they would be systematic about finding all the ways, but I did think that eventually each group would have all the combinations on their cards. The students went right to work, and most groups were able to make six different combinations.

Then I asked for a volunteer to share a combination from the group. One student said, "Three blue and two red, for three boys and two girls." We connected the cubes as a class, first three blue and then two red, to illustrate the student's thinking, and I wrote $5 = 3B + 2G$ on the board and said, "So you have three boys and two girls for the table. Does this number sentence show what you did?"

I asked for a different group to share and another student said her group used four boys and one girl. We connected four blue cubes and then added the one red cube. I wrote $5 = 4B + 1G$ directly above the previous equation. (I was purposefully putting these in order. If the next equation was $5 = 2B + 3G$, I would have written it a space further below the two other equations.) Eventually, we had all six equations on the board and we read them one by one. With each train of cubes completed, beginning with the five blue cubes, I displayed them for all to see:

B	B	B	B	B
B	B	B	B	G
B	B	B	G	G
B	B	G	G	G
B	G	G	G	G
G	G	G	G	G

Now we had a visual arrangement of all the combinations in order. Students were able to notice the stair-step pattern and said, "The numbers go up and the colors go up each time."

$5 = 5B + 0G$

$5 = 4B + 1G$

$5 = 3B + 2G$

$5 = 2B + 3G$

$5 = 1B + 4G$

$5 = 0B + 5G$

One other thing happened during this lesson that I think was helpful for student understanding: One student, Angela, noted, "The five is at the beginning."

I responded with, "We can read this equation as *five is equal to or the same as four blue and one red.*" Equations with one number on the left and an operator on the right—such as $5 = 2 + 3$, to record the number of boys and girls sitting at one of the tables—allow students to understand equations can show in various ways that quantities on both sides have the same value. I then asked if someone would like to read another of our equations and called on a volunteer.

We need to revisit these ideas with other numbers, determining all the ways we can make a given sum with two addends. I think the context of the problem made all the difference in the students being able to understand it. Having the cubes in two colors to represent the two addends also helped the children *see* that various equations are equivalent, but different.

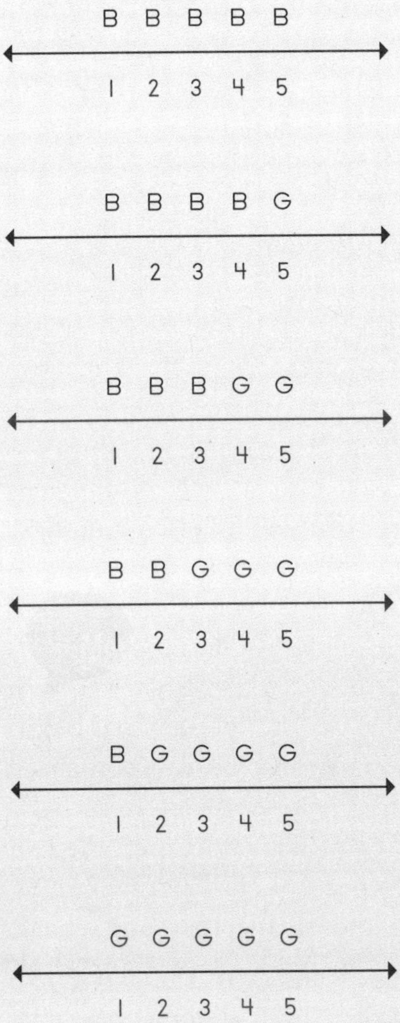

Lesson 9 Solving Comparison Problems

Time

Up to 30 minutes per problem

Materials

More and Less Visual (Reproducible 11)

Story Problems: Solving Comparison Problems (Reproducible 12), each problem on a separate sheet of paper, 1 set (6 problems) per student

counters, books, dimes (as needed to help students solve problems)

Overview

The story problems in this group of lessons encourage students to use an open number line to record abstract and quantitative thinking about comparisons. In the first group of story problems, the goal is to find out how many less or how many more there are when two amounts are compared. There is no *action* of *joining* or *separating*, as in the previous three lessons (Lesson 6, *Solving Story Problems Involving Missing Numbers at the End*; Lesson 7, *Solving Story Problems Involving Missing Numbers in the Middle*; and Lesson 8, *Solving Story Problems Involving Missing Numbers at the Beginning*).

The difference between the two amounts in comparison problems is not present in the situation physically and must be conceptualized or constructed in a representation. In the representation, the extra is shown when the two amounts are matched up. This "extra" amount, when added to the smaller amount, equals the larger amount. This kind of comparison requires looking at both amounts at the same time and making note of the differences. When young children hear a story problem that begins with "Jose has three more crayons than Luis" (and then it goes on to explain that Luis has five crayons), they hear that Jose has three crayons but don't "hear" the part of the story problem where Jose has more than Luis, or the idea of needing to compare how one has more than the other. Students may need many opportunities to listen and hear all the aspects of the situation in order to make sense of the entire comparison. Further, it is often that students can make better sense of "how many more there are" than "how many less or fewer there are" in a given situation. Perhaps young children are just more interested in finding out "how many more" than "how many less"?

In the second group of comparison problems, the goal is to find out how many are in one group when given information about another group that has fewer or more items. These kinds of story problems also do not involve *joining* or *separating*. Students use the number line to compare the numbers, and the space between the numbers represents the difference.

For these reasons, each of the following story problems is complex. One problem may take as long as thirty minutes to read, analyze, discuss, model, represent, and, eventually, solve. This lesson is most appropriate for first and second graders.

Related Lessons

- L-6 Solving Story Problems Involving Missing Numbers at the End
- L-7 Solving Story Problems Involving Missing Numbers in the Middle
- L-8 Solving Story Problems Involving Missing Number at the Beginning

Key Questions

- Who would like to explain what is happening in this story problem using your own words?
- What do these students have?
- What do we want to know?

Teaching Tip

Why Blanks for Numbers?

When introducing story problems, it's recommended to use blanks for the numbers at first, so students do not try to figure out an answer right away. Rather, you want them to focus on understanding what's happening.

Technology Tip

Using an Interactive Whiteboard

If an interactive whiteboard is available, display the problem first with blanks for the numbers. Then, use the pen tool to write in the missing numbers after students have an understanding of what is happening in the problem.

Teaching Directions

Problem One: *Dimes*

1. Write the following story problem where everyone in the class can see it. Consider using your students' names in the problem. Include blanks instead of numbers, as shown here.

   Problem One: Dimes

   Student A has __ dimes. Student B has __ dimes. How many more dimes does Student B have than Student A? If they pool their money, how much money do they have? What do you think they will buy?

2. Read the problem together with students. This can be done with the entire class or in small groups. The paper with the problem and the counters can be made available to students after they have had an opportunity to make sense of the problem. You can support the students in understanding what the problem is asking by using the series of prompts that follow.

3. Use the following statements to facilitate a productive discussion about the story problem:

   - "Tell the story problem in your own words."
   - "Share a different way to explain what is happening in the story."
   - "Tell your shoulder partner what this problem is all about."
   - "What does it mean to pool your money?"

4. Now insert numbers in the story problem as follows:

   Problem One: Dimes

   Student A has 2 dimes. Student B has 5 dimes. How many more dimes does Student B have than Student A? If they pool their money, how much money do they have?

5. Reread the story problem with students. Ask the following questions:

   - "Now that we know how many dimes are in the problem, does someone have less? Does someone have more?"
   - "What do we want to find out?"
   - "How can we explain what is happening?"

6. Have students solve the problem. You could suggest using dimes (or counters to represent dimes) to act out the scenario.

7. Ask a student volunteer to explain how they solved the problem. Record what the student says using an open number line.

Example of Student Thinking and Teacher Recording

Erin: "One of them has two dimes and the other one has five dimes, so one has more than the other."

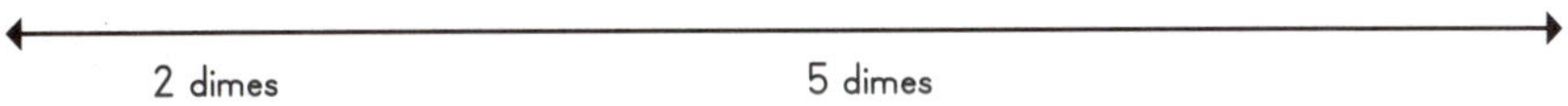

Teacher: "How many more is five dimes?"

Erin: "She has three more."

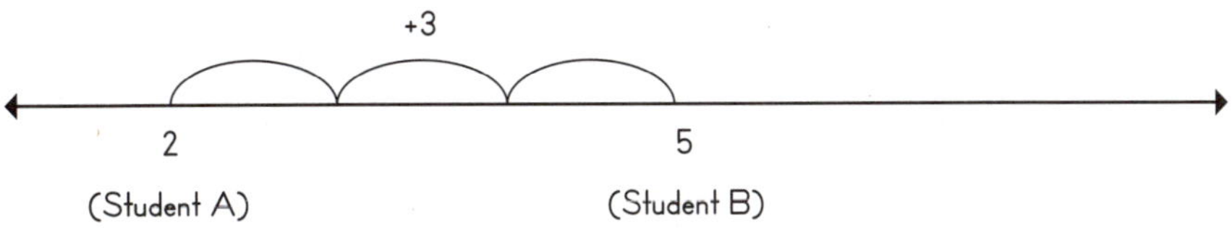

$2 + 3 = 5$

Recording Erin's thinking on an open number line.

Teaching Tip

Why These Numbers?

Smaller numbers have been chosen for these story problems during the whole-class introduction so that students can model what's happening more easily using representational objects, which help them determine the relationship/difference. Students count out a set of objects for the first group, then count out a set for the second group. They then match them up to see if there are any extras that do not have a partner. Students learn that even if one group looks as if it has more objects, matching or counting may reveal a different result.

8. Ask the student volunteer whether your recording (drawing) represents accurately what they said. If the student says it does not, ask the student to repeat their thinking as you change the recording accordingly.
9. Ask a student who has used counters to solve the problem to share their thinking. Record what the student says.

Example of Student Thinking and Teacher Recording

Martin shows his thinking using counters:

The teacher shows and records the comparison:

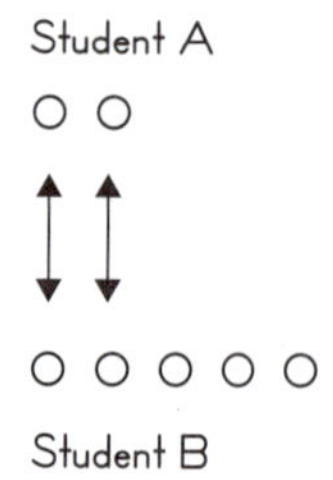

10. Choose one more student volunteer to share their thinking, and record what the student says.

Example of Student Thinking and Teacher Recording

Jenny: "The first one has two dimes and we know the other has more than that. If she gets three more she will have the same."

Teacher: [Draws something like the following.] "What would belong in the part with the '?' if this was a whole-part-part diagram?"

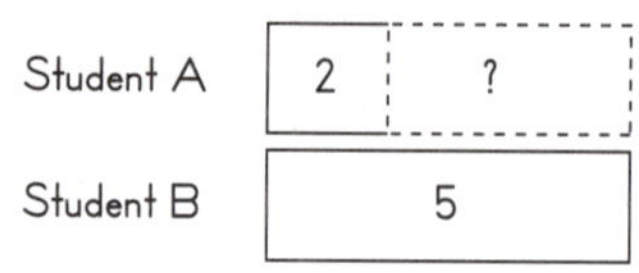

11. Ask students to explain how they thought about the money Student A and Student B had. Encourage them to use counters (or dimes!) to act out the situation.

12. For second graders, extend the problem further by asking, "What is the value of the difference if I know that one dime is worth ten cents?"

Examples of Student Thinking

Chelsea: "Three dimes are thirty cents."

Teacher: "How did you know this?"

Chelsea: "Each dime is ten cents."

Problem Two: *New Books*

On another day, propose the second problem, *New Books*.

13. Write the story problem where everyone in the class can see it. Include blanks instead of numbers, as shown here. Consider using a teacher's name and principal's name in the problem.

Problem Two: New Books

The teacher has ____ new books from the book fair. The principal has ____ new books from the book fair. How many fewer books does the teacher have than the principal?

14. Read the problem together with students. This could be done in small groups or as a whole class.

15. Use the following statements to facilitate a productive discussion about the story problem:

- "Tell the story problem in your own words."
- "Share a different way to explain what is happening in the story."
- "Tell your shoulder partner what this problem is all about."

Math Matters!

Two-Part Problems

The *Dimes* Problem is an example of a two-part problem. The first part requires students to figure out the difference and the second part requires students to determine the value of the difference. When children are exposed to complicated problems, having the opportunity to explain their thinking with the support of a group increases their ability to solve these kinds of problems on their own.

16. Now insert numbers in the story problem as follows:

Problem Two: New Books

The teacher has 2 new books from the book fair. The principal has 5 new books from the book fair. How many fewer books does the teacher have than the principal?

17. Reread the story problem with students. Ask the following questions:

- "Now that we know how many books are in the problem, does one of the people have more books? Does someone have less books?"
- "What do we want to find out?"
- "Who has a way to explain what to do?"

18. Have students solve the problem. You can suggest the problem be acted out using students to represent the principal and the teacher and also using books as counters.

19. Ask a student volunteer to explain how they solved the problem. Record what the student says using an open number line.

Teaching Tip

Things to Keep in Mind with Story Problems

1. Change the story problem as needed to reflect the interests, experiences, and names of students.
2. Present only one or two stories at a sitting.
3. Check to ensure that children can identify groups with more or less by displaying the More and Less Visual (Reproducible 11).
4. Instead of telling children how to do the work ("you need to subtract"), ask them to explain the relationship of the numbers in the problem or use objects to represent the numbers.
5. Include additional information that may help to make the problem more realistic. It also may be helpful to include possibly nonessential information to make the problem more engaging or to encourage children to evaluate the problem for information important to finding a solution.

Example of Student Thinking and Teacher Recording

Olivia: "The principal has more than the teacher."

Mona: "The teacher has less than the principal."

Chelsea: "The teacher has only two and the principal has five, so that's three less."

Teacher: "So you are saying that less and fewer are the same idea? How many less would the principal have if the principal had the same number as the teacher?"

Maggie: "If the principal had three less, the principal would have the same as the teacher."

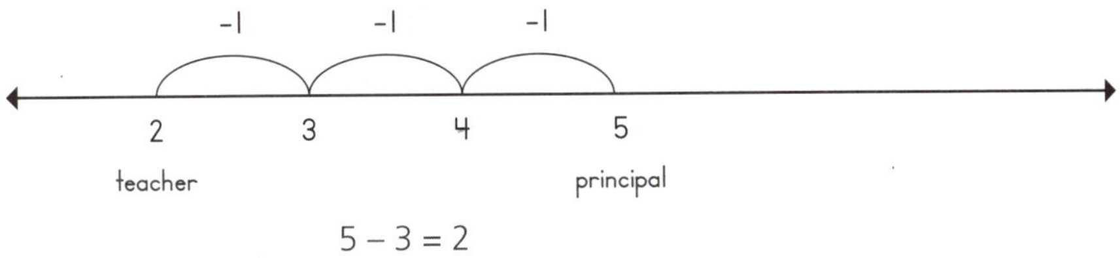

Recording students' thinking on an open number line.

20. Ask the student volunteer whether your recording (drawing) represents accurately what they said. If the student says it does not, ask the student to repeat their thinking as you change the recording accordingly.

21. Ask students to explain how they know the difference between the two sets of books. Encourage them to use counters (or books!) to act out what is happening in the story.

22. Have a student volunteer share their thinking aloud. Draw the books as the student explains.

Example of Student Thinking and Teacher Recording

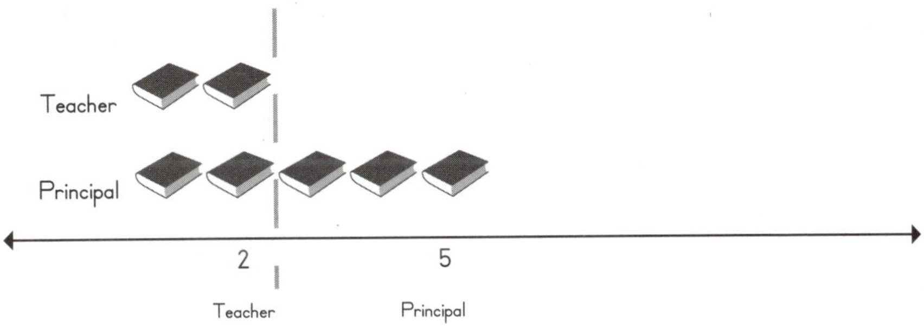

23. Ask the student volunteer whether your recording (drawing) represents accurately what they said. If the student says it does not, ask the student to repeat their thinking as you change the recording accordingly. Wrap up the problem by asking, "What book do you want to buy from the book fair? Do you have a favorite?"

Problem Three: *Marbles*

On another day, you can continue with the third problem, *Marbles.*

24. Write the story problem where everyone in the class can see it. Include blanks instead of numbers, as shown here, and consider using your students' names in the problem.

    Problem Three: Marbles

    Student A has __ more marbles than Student B. Student B has __ marbles. How many marbles does Student A have? How many do they have if they put their marbles together?

25. Read the problem together with students.
26. Use the following statements to facilitate a productive discussion about the story problem:

    - "Tell the story problem in your own words."
    - "Share a different way to explain what is happening in the story."
    - "Tell your shoulder partner what this problem is all about."

27. Now insert numbers in the story problem as follows:

    Problem Three: Marbles

    Student A has 3 more marbles than Student B. Student B has 2 marbles. How many marbles does Student A have? How many do they have if they put their marbles together?

28. Reread the story problem with students. Ask the following questions:

    - "Now that we know about some of the marbles in the problem, does one person have less or more than the other person?" (Use your students' names in the problems and discussion.)
    - "What do we want to find out?"
    - "Who has a way to explain what to do?"

Math Matters!

Two-Part Problems

The *Marbles* Problem is an example of a two-part problem. At first, students need to find out what the amounts are when the difference is known. The second part requires students to combine the two amounts to determine the total number. When children are exposed to complicated problems, having the opportunity to explain their thinking with the support of a group increases their ability to solve these kinds of problems on their own.

29. Have students solve the problem. Encourage students to role-play the problem and use counters to show their solution strategies.

30. Ask a student volunteer to explain how they solved the first question: How many marbles does Student A have? Record what the student says using an open number line or other visual representation.

Example of Student Thinking and Teacher Recording

Mario: "Student A has three more than B. So, since B has two, A has three more."

Mario continues by showing the following.

Mario: "Here's what B has."

Mario: "Here's what A has."

The teacher helps Mario line up the counters, so they can be compared.

Teacher: "How should I show these numbers of marbles on the number line?"

Mario: "You start with two where B is and jump three more and get to five where A is."

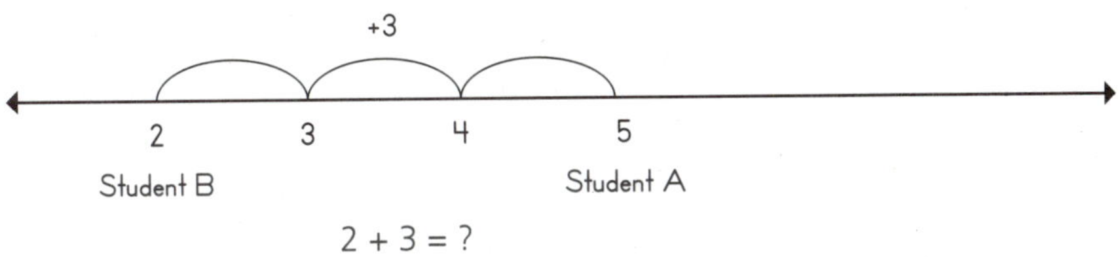

Recording Mario's thinking on an open number line.

At this point, the teacher asks a volunteer to explain how the equation connects to the number line.

31. Now ask how students solved the second question: How many marbles do Students A and B have if they put their marbles together? There should be a great deal of conversation and dramatic action to ensure students make sense of and role-play this complex problem.

Example of Student Thinking and Teacher Recording

Mario: "Here's what B has."

Mario: "Here's what A has."

Mario: "So, they both have this many."

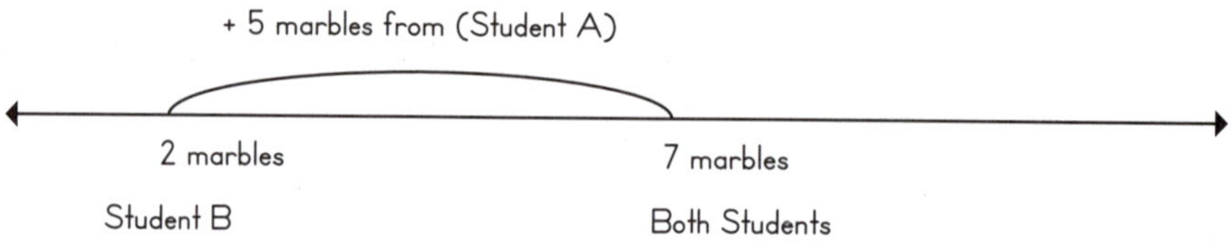

Recording Mario's thinking on an open number line.

32. What game do you think these two students should play with their marbles? This mini-conversation brings a real-world connection to the problem they have solved and engages students in thinking about doing something with the "answers."

33. After students have solved the problems, hold a whole-class discussion and encourage students to share their strategies as you record their thinking.

Teaching Tip

Extending Student Thinking

On another day, you can distribute individual or several story problems from Story Problems: Solving Comparison Problems (Reproducible 12). The numbers have been changed slightly from the ones used in the teaching directions. You may wish to change the context to match the interests of your class. If you do this, make sure you keep the structure the same. Have students work on the problems independently or in pairs.

Another option is to have students keep the problems in their math folder and work on them when they finish other math classwork.

Teacher Reflection

The Complexity of Teaching Different Types of Problems

Consider these two problems:

1. There are five boys and eight hats. How many more hats than boys are there?
2. There are five boys and eight hats. If each boy puts on a hat, how many hats are there left over?

Carpenter, Fennema, and Franke (1994, 10) found that the second problem is easier for students because the action is more explicit. (Chapin & Johnson 2006, 59)

When I first read about the two previous problems, I was reminded of how important it is for my students to act out a story problem so they can get a sense of what is happening in the problem. In the first problem, my students just counted the things in the problem—eight hats and five boys—and got thirteen things. The second part of the problem—how many more hats than boys—did not relate to their counting of things. When I presented the second problem, even though we did not have any hats in the room, several children grabbed pieces of paper and put them on the heads of some of the children who had been designated "boys" in the problem. The problem situation became real to them with their actions, even though some of the girls were playing the parts of the boys in the problem!

The magnitude of the numbers in a problem also affects its level of difficulty. The following problems are both separate problems in which the change is unknown, but the second one is more difficult for young students because the numbers are larger:

1. Liz had 12 pennies. She gave some pennies to Caitlin. Now she has 8 pennies. How many pennies did Liz give Caitlin?
2. Liz had 45 pennies. She gave some pennies to Caitlin. Now she has 28 pennies. How many pennies did Liz give Caitlin?

However, after students solve many separate/amount-of-change-unknown problems that involve small quantities, they are more likely to be able to answer similar problems with larger quantities. (Chapin & Johnson 2006, 59)

So, one of the strategies that I often suggest to my students is to change the numbers in the problem to see if they can get an idea of what the problem is about. The previous problem could be made even more comprehensible as follows:

> Liz has 5 pennies. She gives some pennies to Caitlin. Now she has 3 pennies. How many pennies did Liz give to Caitlin?

When we do this, students seem to be able to role-play the parts of Liz and Caitlin, with a few coins or counters, and have a sense that Liz started off with an amount and ended up with less. When the students see the relationships, it is much easier for them to connect to the original problems. I have found that sometimes children are afraid of larger numbers and shut down. I want them to approach challenging problems with strategies that give them support and confidence.

> Students' abilities to translate the words in a problem to an operation that represents the relationships presented by the words takes time and many experiences. Teachers need to present all problem types throughout the school year so that students have the opportunity to develop meaning and fluency for word problems. (Chapin & Johnson 2006, 59)

When I read this, I was reminded that I need to present different kinds of problems, different structures, and different formats, with the emphasis on making sense of the problem, acting out the problem, and persevering until we are satisfied we are done. This means I won't put all join problems together in one week and all separate problems together in another week; rather, I mix them up. For children to be successful problem solvers, they need to grapple with all different problem types and make visual representations of what is happening in the problem. They won't be able to *remember* that a *how-many-more* problem can be *solved by subtracting*. They will, however, be able to compare two groups and match them up to see how many extras there are beyond the smaller set.

Estimating and Gathering Data

Lesson 10

Overview

In this lesson, students estimate the number of items in a "mystery jar" and record their estimates on an open number path. The students determine how many are in the jar by counting by ones, twos, or fives. Then the students compare their estimates to the actual count. Students build number sense as they learn the actual count, identify the close estimates, and revise their estimates. The task provides repeated opportunities to record one's thinking as well as build independent thinking. It's recommended that this lesson be done once a week for the entire school year to establish a routine and ensure the development of number sense and estimation skills over time.

Time

15–20 minutes, once a week

Materials

strip of paper cash register tape (about 10 feet in length)

3-inch-by-3-inch red paper squares, 25

3-inch-by-3-inch yellow paper squares, 25

glass jars, 2 of the same size, 1 with 30–40 cubes (considered the "mystery jar"), the other with 10–20 cubes (considered the "reference jar")

Related Games and Lessons

- G-1 Mystery Number (Version 1)
- G-2 Mystery Number (Version 2)
- G-3 Mystery Number (Version 3)
- G-4 Race to 50
- G-6 101 and Out!
- G-9 Adding Nines, Tens, and Elevens
- L-3 Building an Open Number Path

Key Questions

- What is your estimate?
- Why do you think your estimate is reasonable?
- How do you know whether your estimate is close to the actual count?

Teaching Directions

1. Before beginning the lesson, prepare an open number path, per the materials list, with fifty segments marked by red and yellow squares (the color should alternate every five squares). Display the number path on a classroom wall at students' eye level.
2. Gather students together, in pairs, where they can see and access the number path easily.
3. Explain the activity. "Today we are going to estimate how many cubes are in our mystery jar and record our estimates on our number path. Talk with your shoulder partner and decide on an estimate that you think is close."
4. Show the class the mystery jar (the jar containing thirty to forty cubes). Also share the reference jar (the jar with ten to twenty cubes) and explain to students that they can use it to help with their estimate. Share the quantity of cubes within the reference jar. Say, for example, "This reference jar has ten cubes in it. Use this information to help you choose your own estimate for the cubes in the mystery jar."
5. Ask each pair of students to write their estimate on a sticky note.
6. Ask each pair of students to "stick" their estimate where appropriate above the number path. Students can have the same estimate as other classmates (this means they agree with each other that this is a reasonable estimate). When they do, make sure they put their sticky note above the previous sticky note, not over it (thus building a column of sticky notes above the number path).
7. Now ask students, "How did you know where to put your sticky note?"

Teaching Tip

The Benefits of Partner Work

Estimating with partners allows students a chance to discuss their estimates and reduces the number of random or highly unlikely "guesses." In this activity, student names are not associated with the estimations, so "winning" or "being correct" is not the focus. Students eventually see that a close or reasonable estimate is the objective for the task.

Teaching Tip

The Reference Jar

Sharing a reference jar encourages students to use information about a known quantity to make their estimates.

Math Matters!

Unitizing

When students, without counting one by one, are able to put their sticky note estimate of 21 right above 21 on the open number path because it is next to 20 (the last square in the fourth group of five), we know they are making sense of a group of ten or groups of five.

8. Now share sentence frames such as the following:

   Our estimate is greater than ________.

   We think our estimate is close because _________.

   We estimate _________ because _____________.

   Model the use of one of the frames by pointing to one of the estimates and saying, "We think our estimate of 20 is close because it looks like there are about ten more in the estimate jar than in the reference jar, which has ten."

9. Ask students to choose one of the sentence frames and turn and tell their shoulder partner about their estimate.

10. Call on a student to describe their own estimate to the rest of the class using one of the sentence frames. Repeat, calling on several other students.

11. When all the sticky notes are posted, you will have a graph that displays the estimates for the mystery jar. Ask students to describe the range and frequency of the estimates:

    "Which is our greatest estimate?"

    "Which estimate is least?"

    "What do you notice about our estimates so far?"

Differentiating Your Instruction

Creating Accessibility for All Students

Sentence frames support all learners, but especially those learners for whom English is their second language. Visual references also help all students. Consider having a math word wall with words and illustrations of key ideas such as these (also available as Reproducible 2):

5 is greater than 3

4 is less than 7

5 is equal to 5

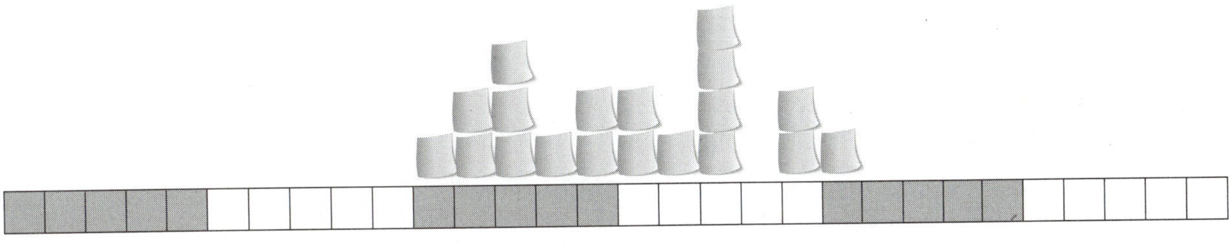

Estimate graph.

Examples of Student Thinking

Ramona: "The greatest estimate is twenty-one."

Eva: "The smallest estimate is eleven."

Teaching Tip

Revising an Estimate

Revising an estimate is a chance for students to reconsider what they guessed based on new information. Early, in the beginning weeks of this activity, consider asking students to think about revising their estimates when the actual count is almost complete. In this way, they will notice how close their revision could be to the actual count. Later, consider revising when the class is about halfway through the count to support number sense reasoning.

A Child's Mind . . .

Estimations

Watch to see whether students make use of the reference jar or the revision opportunity when they are involved in estimation. As students become more comfortable with the task and the quantities, their estimations become more reasonable.

Jason: "There's more on eighteen."

Martin: "There's none on nineteen."

Teacher: "So the range is between eleven and twenty-one."

12. Empty the contents of the mystery jar onto a table. Count the cubes out loud, either placing them back in the jar one by one or counting them in groups of twos or fives.
13. Stop about halfway through the counting process and ask, "Would any pairs like to revise their estimates?" If pairs decide to do this, let them replace their sticky note estimate with a new one and ask them, "Why did you decide to revise your estimate?"
14. Examine the range of the estimates and make note of any changes in the frequencies.

Examples of Student Observations

Teacher: "Now our range is from thirteen to nineteen. What else do you notice about our graph?"

Serena: "Now more guesses are on fourteen."

Arturo: "Someone changed to nineteen now."

Revised estimate graph.

15. Continue counting the cubes back into the jar until they are all replaced and an actual count has been announced. Ask, "Is there a sticky note for this amount on the graph?" If so, have a student go up and put a star on the corresponding note or column of notes.

16. Ask, "Are there any estimates that are close to our actual count? How do you know they are close?" Encourage students to come up and count the diference between the "close" estimates and the actual count. If you do not have sticky notes posted near the actual count, ask students to suggest a number that *would have been close*. In this way, they can practice thinking about numbers that are close to the actual count.

Teaching Tip

How Do You Know This Estimate Is Close to the Actual Count?

There are other ways to determine "closeness" besides looking at the paper squares or thinking about the numbers. Students can measure with their hands the distance between the two numbers. Students can also count up to the actual number from the estimate and see (and hear) that it doesn't take too many numbers to reach the actual count. Ask them to explain how they know this estimate is close to the actual count.

Differentiating Your Instruction

Choosing the Range for the Number Line

When you begin estimation tasks with younger children, use quantities from one to twenty or even one to ten. After they become more familiar with estimations, older children can make sense of a range to fifty or more. Use the same-size jar each time, and provide a matching reference jar with ten, twenty, or thirty counted items so students can use information about a known quantity to make their estimates. When the jar contains more than one hundred items, students are more apt to guess random numbers than make thoughtful estimates. Use items that are roughly the same size on subsequent days, so students can make use of estimates that were close on previous days.

Extend Their Learning!

A Classroom Routine

Make this activity a regular part of students' classroom routine—for example, as part of their morning check-in at the beginning of the day. As students enter the classroom and record the lunch count and attendance, have them also record their estimates. Later during the day, hold a class discussion to address their estimated data and their growing estimation skills.

Make Your Own Mystery Jar

Collect pairs of straight-sided glass jars (olive and jam jars work well because of their smaller size) and store them in plastic bins. Encourage students to make their own mystery jars for their classmates to estimate and count. Offer an assortment of possible counters: lima beans, counting bears, paper clips, bottle caps, bread tags, buttons, shells, pebbles, marbles, pennies, pom-poms, and so on. Make this an activity that students can do when they arrive and settle into the classroom. Consider giving students the opportunity to share their mystery jar with the whole class. After they're familiar with the activity's procedures, encourage students to take charge and lead an estimating activity.

Teacher Reflection

Estimates and Guesses

I've learned that its helpful to have a conversation with students about the distinction between an *estimate* and a *guess*. I bring up the idea that when we guess, we do not necessarily have a numerical reason for the guess. Sometimes we may guess our favorite number, our birthday number, or a number that is easy for us to write. When we estimate, we use information of some kind to support our thinking. If we can compare the estimate to something that we know for sure, our estimate is more reasoned than a guess.

I knew that estimation was becoming a useful strategy for my students when Jason said (after noticing that the reference jar, which contained ten pecans, was about one-third full), "The mystery jar is almost full and so we knew that it has to be like ten and ten and ten, so we guessed thirty." As he described this, his partner showed the two-inch segments on the side of the jar with his thumb and forefinger.

With experience using the same-size jar over time, my students have come to realize that, when filled, the mystery jar holds about thirty pecans in their shells. When presented with the same jar filled with walnuts, students then reason that because walnuts are larger than pecans, there might be fewer walnuts in the jar, perhaps twenty.

I'm confident that children develop number sense when they have experiences that contribute to their reasoning abilities; I worry, though, about the children who continue to make wild guesses. Maybe these quantities are too big for them? I think I need to find ways for these children to deal with fewer items. Maybe they could make their own estimate jars for the rest of the class and present them to the class as a challenge.

Games Using Number Paths and Number Lines

(Continued)

Why these math games?

The games in this section provide excellent opportunities for students, working in pairs or small groups, to use number paths and number lines to reinforce concepts and skills introduced in the class lessons (Lessons 1 through 10). Games that are enjoyable for students are more likely to be played often, giving students the repeated experience they need with concepts like number comparison, addition, subtraction, and place value. Repetition is important with games; each time students play, they benefit from a game in a slightly different way. The first couple of times playing a game, the focus tends to be on the logistics (how to move, what to do on your turn, and so on). Subsequent play allows students to interact with the game's concepts in a deeper way, focusing on the big ideas.

The first games in this section, *Mystery Number, Versions 1–3* (G-1–G-3) and *Race to 50* (G-4), deal with foundational concepts such as the order of numbers on the number line and being able to think about numbers in chunks by unitizing for five or ten. Next, the following set of games, *The Game of Pig on a Number Line* (G-5), *101 and Out!* (G-6), and *Hot Lava Bridge* (G-7) involve making decisions using place value while dealing with strategy and chance. Then, the collection of games dealing with differences and comparisons, *The Larger Difference Game* (G-8), *The Smaller Difference* (G-9), and *Adding Nines, Tens, and Elevens* (G-10), provide opportunities for students to make sense of greater than and less than expressions. Finally, *Get to the Target, Versions 1–3* (G-11–G-13) and *Race to 1,000* (G-14) are games that use all of the skills students will need to make sense of adding and subtracting without encountering the common errors associated with the traditional regrouping algorithm.

When should I teach with math games?

Finding time to give students repeated opportunities to play games can be challenging but rewarding. After you've taken time to introduce students to a particular game, have played the game during math time, and have had an opportunity to discuss the content and strategies, you can use that game repeatedly in stations, or when there is extra time during the day. This extra time might be in the mornings, when students are arriving and putting away their backpacks (during your *settle-in time*), during rainy-day recess, or during transition times (if there is time left during lunch, for example).

Do I need to prepare materials in advance for these math games?

Some teachers prepare file folders containing the recording sheets for the games and include sandwich or freezer bags containing counters or game cards. Other teachers provide these items in "May Do" bins for students to use when they have finished all their work and need something to do. When students are familiar with the games, they usually do not have difficulty in going to a place in the classroom where game supplies are located to select the materials they need.

Number Lines

Many of the games require the use of an open number line. Once familiar with the open number line, children can easily draw them on their own. Sentence strips, cash register tape, and blank paper can be stored in bins or on shelves, so they are readily available for students to use in the construction of number lines. In some cases, the games use the red and yellow blank squares number path (arranged in groups of five). The number path with numerals from 1–10 or 1–20 can be teacher-made on sentence strips or student-made with numbered paper squares. These can be tools that are regularly posted on the wall at child level in the classroom.

How do I manage my classroom during math games?

It's important to manage students during game play, and there are many techniques for doing so. It is your role, as the teacher, to set clear expectations and to supervise students while they are playing games. Many of the issues that arise when students are playing games can be avoided via careful preplanning. The following paragraphs provide a few helpful tips.

Assigning Partners

Assigning partners can be awkward for students, especially when they are assigned to play with someone whom they do not consider to be a friend. Prepare students for assigned partners by setting clear expectations. Make it a goal from the beginning that students will have many different partners during the school year. State explicitly that it is inappropriate to show expressions of dismay or to say unkind words. Discuss disciplinary actions as necessary. In addition, let students know the length of time they will be expected to work with their partners, such as thirty minutes or one week.

Decide on your goals for pairing students. Do you want one student to support and help another? Do you want two students to work together on the same level and push each other? There are many benefits of random grouping for partner and small-group work. When students are assigned a partner or group randomly by selecting a playing card or by pulling fair sticks, they know they are entering a working relationship, not a lifetime pairing. They are expected to focus on doing the work well together. When assigning partners, it is important to let students know that no matter who they have as a partner, you expect them to listen to one another and try their best to complete the task assigned to them in a timely manner. Set expectations and make it clear that they will have many different partners during the school year. State explicitly that it is inappropriate to use unkind words or actions. Discuss disciplinary consequences as necessary. In addition, let students know the length of time they will be expected to work with their partner or group, such as thirty minutes or two days. When you let them know their partner has been chosen at random, they see the method is fair and you expect them to do their best to live up to your expectations.

Letting Students Choose Partners

After the year is well under way and you have clearly established routines for playing games, consider allowing students to choose their partner. One structured way to start the process is to ask a few students as they arrive in the morning, "Whom would you like to play a game with during math?" Start a list of those who choose partners (and include their corresponding partners). For those who do not choose a partner, assign them one and include them on the list. During math, read or post the list of partnerships without stating who picked their partners. This way, students won't feel like they were picked last.

In addition, when you give students the opportunity to choose partners, it's helpful to discuss with the class that sometimes best friends do not make the best game partners. Encouraging students to choose a game partner with whom they work well helps students become responsible for their own learning. Once again, setting clear expectations is crucial; explain that if students choose someone and end up not following directions or not working well together, you will choose new partners for them.

Playing at Desks Versus the Floor

At the beginning of the year, as you are establishing routines for playing games, it is easier if students play all games at their desks. If they are using a single or double number path or number line, ask them to sit next to their math partner so that the numbers will be right side up for both of them. They will benefit from being able to see the numbers their partner is marking. If the game requires students to record something as they play, working at a desk is better for this task than on the floor.

After routines are in place and students are demonstrating appropriate game-playing behavior, you might choose to move them to the floor. Students are often more relaxed sitting on the floor, and materials may be easier for students to use when they are on the floor. When students play games on the floor, have them use clipboards or their math journals as "desks" for their worksheets while recording the results of their game. Sitting on the floor provides students with a more flexible seating arrangement as well; they can either sit face-to-face or elbow-to-elbow. If students begin to act silly or are not following directions while playing on the floor, warn them that they will have to move back to their desks unless they behave appropriately. Be sure to follow through after your warning.

Managing the Use of Game Pieces: Counters, Game Cards, and Dice

Clear organization of a variety of materials needed for games is crucial. Materials need to be readily accessible to students. Keep manipulatives in clear containers or plastic bags that are placed low enough for students to reach. Consider keeping the reproducibles for games as well as the materials all together in large plastic bags.

Regarding clean up, it is typical in any classroom to find game pieces left over on the floor from a previously played game. During the beginning of the year, explain your procedures for cleanup, including having students check around them, under them, and nearby for any stray materials. Be sure to allow ample time for students to practice these procedures early in the year while you monitor them, so that later in the year they will know your expectations for cleanup without your direct supervision.

Young students are notorious for following teachers around the room, letting them know they found a counter, a tile, or a game marker. To help build autonomy, at the beginning of the year, hold up a found item then ask students to be problem solvers; encourage them to think quietly about where the item goes. Give students time to share their thoughts with their partners, and then hold a class discussion about what to do when misplaced or lost items are found. This type of discussion encourages students to take ownership of their classroom and put away these items on their own.

Setting up another container or large plastic bag and labeling it *I'm lost!* for lost items helps students know what to do when they find a misplaced game piece but are unsure of where it belongs. At the end of the day or week, the teacher (or a resourceful and aware student) can place these lost items in their proper place.

Why is it important to model math games?

For young students, modeling how to play a math game is critical for the success they will have while playing it. It is usually best to seat students on the floor while modeling a game. Have students sit on the perimeter of the whole-group area and place the game materials in the center. This arrangement ensures that each student has a chance to see and hear what is going on while also being in close proximity to you. In addition, you can tell quickly who is and who is not paying attention and can redirect behavior as necessary.

It helps to assign places for students to sit in the whole-group area; doing so alleviates students' desire to run to the front of the area or to save a seat for a friend. Also, consider placing students strategically, specifically those who would benefit from sitting in the front of the group or closer to you.

Sometimes teachers choose to introduce games to younger students in a small group rotation. For example, it is often easier for kindergarten children to understand how a game works if they do not have to wait long for their own turn. In this setting, they also can touch their own game pieces when it is their turn because of the closeness in a small table group.

There is more to modeling a game than simply explaining the instructions and walking through a pretend game. It's important to teach students explicitly the behavior you expect them to display while playing games in the classroom. For some students, this may be the first time they have played a card or a dice game and they may not know good game etiquette. Emphasize the following three behaviors when modeling: waiting patiently for your turn, passing game pieces to your partner appropriately, and winning or losing gracefully. It is also important for students to hear others' thinking about the game. While modeling the game, ask the key questions included with each game, so that important mathematical ideas are discussed.

Modeling How to Wait Patiently for Your Turn

To model waiting patiently for your turn, remind students of what their role is while they wait. They might need to be checking their partner's work or their own work from previous rounds. Help them see that waiting is not a passive activity when playing math games but is an active one. Let them know, also, that waiting patiently does not mean telling their partner repeatedly what to do or giving their partner the answer. Emphasize that such actions take the learning away from their partner. On the other hand, tell students that if their partner asks for help, they may help.

Modeling How to Pass Game Pieces Appropriately to Your Partner

Passing dice or cards to a partner may seem trivial but can turn quickly into an argument. Demonstrate for students how to wait politely for your partner to pass the dice when their turn is over, rather than grabbing to begin your turn. Remind students that they should be watching their partner take a turn and checking that they are following all the procedures of the game. Consistent teacher modeling of the appropriate way to pass game pieces will make this act second nature to students.

Modeling How to Win or Lose Gracefully

Learning how to win or lose gracefully is an important life skill for children. Whether playing math games at school or board games at home, students need to learn this skill, and modeling in the classroom is an important part of that learning. At the beginning of a game, explain that someone will win and someone will not win *this time*.

Explain to students that they have a choice about how to react at the end of a game. Whether they win or lose, their first statement should be, "Good game!" Remind them that everyone loses sometimes and, although it is okay to feel disappointed, it is not okay to act out that disappointment with unkind words or actions. In the same way, nobody wants to play with someone who gloats at the end of the game they win. (*Gloat* might be a good vocabulary word to introduce!) Tell them that being a gracious winner or loser will ensure that friends will want to play the game with them again. Note: Having students play two-against-two can also help them feel more comfortable during competitive games.

Teaching Tip

Games without Winners

Sometimes students have difficulty understanding that some games don't have a winner or a loser. Games such as *Mystery Number, Versions 1–3* (G-1–G-3) and *Get to the Target, Versions 1–3* (G-11–G-13) are played cooperatively, with the goal being to complete a particular task—in this case, to determine the mystery number or find different ways to get to a target number. Students take turns and work together to complete the task. Be sure to explain that some games are played for the enjoyment of working together, and that during these games, both players are winners at the end.

Teaching Tip

Brainstorming How to Win or Lose Gracefully

Losing a game can seem like a big deal to a young child. It is helpful for them to have some words to say to express their feelings in an appropriate way. Hold a brainstorming session with students about phrases they can say when they win or lose. When students lose they might say, "Oh, bummer. Maybe next time." Then ask students to think of other kind phrases they can say when a game is over. As students volunteer their thoughts, record the phrases on a poster or chart paper. Title the list, "What to Say When a Game Is Over" and post it on the classroom wall. As the year goes on, students can glance up and find words to say if they have trouble coming up with their own. During check-in discussions after games, ask students to report kind words they heard their partner say, and add them to the class list if these are new phrases.

Role-Playing and Observing Positive Game Playing Behaviors

One strategy many teachers use to ensure student understanding and the use of these positive behaviors is to role-play appropriate and inappropriate interactions with students. Students are delighted when the teacher plays the role of the student who *does not* portray the desired actions. The students in the audience provide feedback and suggest corrections for the teacher, stating what the teacher should do better next time and why this is a good idea. This review can be repeated as necessary with small groups, so all students have a chance to practice talking about and using positive game-playing behaviors. If students persist in not following the standards for positive game-playing experiences, sitting to the side and watching others follow the rules of the games will be an instructive alternative activity for them to do. A checklist of these observations can be a useful tool for these student observers. Here is an example of the observer's checklist:

How many times did you see the following? Mark each time with a check.

1. Waiting patiently for a turn: ✓✓✓✓
2. Passing game pieces appropriately: ✓✓✓✓✓✓
3. Winning or losing gracefully: ✓✓

What should I do while students are playing games?

While students are playing games, teachers should observe students, make note of what students write on their recording sheets, ask questions to extend students' learning and to access their understanding, and work with small groups of students who need more help or in which behavior issues or partner disputes arise.

Using Recording Sheets

Recording sheets serve as both formative and summative assessments. Having students record their thinking while playing games is a valuable way to gain insight into students' understandings. Before introducing a recording sheet, make sure students are familiar with the game and have played it several times. Don't forget to model how to complete the recording sheet before handing it out to students. After students have started the game and are using their recording sheets, move around the room and note what students are writing. Ask students questions about what they have recorded. In addition, collect and review students' recording sheets to gain insight for planning next steps for the entire class or small groups of students.

Most teachers are required to indicate levels of student understanding by assigning grades or checking off indicators. Recording sheets can serve as summative assessments when you need to assign a grade. Last, whether you use recording sheets to plan future lessons or indicate what has been learned, you'll find them helpful when discussing students' understandings with parents and administrators.

Mystery Number (Version 1)

Game 1

Overview

During this activity, students try to identify a mystery number on a number path using a series of guesses and clues. The holder of the mystery number gives clues using the words *less* or *more* and *less than* or *greater than*. The range of the guesses is eventually narrowed, supporting students in identifying the number. The examples in this lesson's directions are from a kindergarten class; for higher grades, increase the range of the number path and/or see *Mystery Number* (Version 2).

Related Games

- G-2 Mystery Number (Version 2)
- G-3 Mystery Number (Version 3)
- G-11 Get to the Target (Version 1)
- G-12 Get to the Target (Version 2)
- G-13 Get to the Target (Version 3)

Key Questions

- What is your reason for your guess of ___?
- Why might your guess be a good guess?

Time

10–15 minutes

Materials

number path labeled *1* through *10* or *1* through *20*, with the numbers spaced equally apart (approximately 3 inches), created using paper sentence strips or cash register paper that is approximately 3 inches tall

sticky notes (approximately 3-inch-by-3-inch squares)

Reproducible 2, Greater Than, Less Than, and Equal To Reference Chart (enlarged for classroom display)

Extension

open number path, created using cash register tape (approximately 10 feet)

red and yellow paper squares (approximately 3 inches by 3 inches), 10 of each color, organized in groups of five starting with red

clothespins with arrows on them, like this:

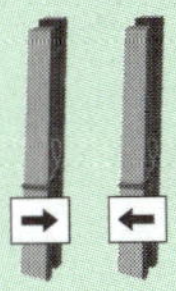

Teaching Directions

Introduce

1. Prepare a number path per the Materials list. Make sure the number path is large enough for everyone in the class to see it. Display it appropriately.

1	2	3	4	5	6	7	8	9	10

Number path labeled *1–10*.

2. Gather students on the rug so that everyone has a good view of the number path. Share with students that they will be playing the game *Mystery Number*. Explain, "I am thinking of a number somewhere on the number path. Your job is to try to guess my number. I will give you clues each time you make a guess."
3. Choose a number. For the purposes of these directions, let's say the number you choose is 7. This is the mystery number; do not share it with students. Consider writing the mystery number on a piece of paper and turning the paper over so students can't see it.
4. Ask for a guess from a volunteer. Record the guess where everyone can see it, then respond to that guess with a clue. For mystery number 7, if a student guesses 5, respond with the clue, "My number is greater than five."
5. Now cover the number 5 on the number path with a sticky note to indicate it has been guessed. Ask, "Could the mystery number be four if I said that it is more than or greater than five?" Gesture to more than five with a hand signal. For example, you

Teaching Tip

Choosing Numbers and Path or Line

When determining the range of numbers to use on a number path, be mindful of students' number knowledge. For instance, kindergarten students are likely to be more comfortable with a number path to 10 at the beginning of the year and could work with number paths to 20 later in the year. Older students are able to work with an open number line and know, for example, that 35 falls "somewhere after the middle" of a 0–50 open number line.

Teaching Tip

***Somewhere* on the Number Path**

Be careful of your language when describing the game. If you say you are thinking of a number between one and ten, students might be confused about whether the number could be one or ten. For this reason, the directions are phrased intentionally to indicate to students that the number *is somewhere on the number path*. This also prompts students to think about the entire number path when making a guess.

could show this by pretending to hold a ball with two hands that is being blown up with air to show more. When you need to gesture less, you can bring your hands closer as if holding a ball that is losing air.

Example of Student Thinking

Ramon: "No, because five is more than four."

Silvia: "It has to be six or seven or eight or nine or ten."

6. Ask, "So, because four is not my number, should we cover it up with a sticky note?" Students should respond with a resounding yes.
7. Ask, "Are there other numbers we could cover up at this time?" Hopefully students see, per the number path, that the numbers 1, 2, and 3 can also be covered.
8. Say, "Who has a second guess? Remember, my number is more than five."

Example of Student Thinking

Bodie: "Is it six?"

Teacher: "My number is more than six."

9. Continue encouraging student guesses, providing more clues based on the guesses and reminding students of the previous clues. For example, if the third guess is 10, say, "My number is more than five. It is also more than six. And my number is less than ten."
10. When students identify the mystery number, review all the clues once more:

    "Is seven more than five?" (Yes.)

    "Is seven more than six?" (Yes.)

    "Is seven less than ten?" (Yes.)

    "You guessed it. Seven was my number."

A Child's Mind . . .

Strategies for Efficient Guessing

Understanding which numbers to guess may not be readily apparent to students until they have played the game many times. One of the most efficient strategies for thinking about which mystery number to guess is to choose a number that is halfway across the range of remaining numbers.

Often, students guess a number that seems random but is important to them, such as their age or birthday. When the range is reduced dramatically and made apparent visually by covering numbers with sticky notes, children begin to notice how quickly they can eliminate the greatest number of possibilities with one guess. This is an example of strategic and logical mathematical learning based on the thinking that is within the child's mind. It cannot be taught effectively to students by just explaining or showing. Until this idea becomes apparent to children, they may guess: "Is it one?" "Is it two?" "Is it three?" and so on.

A Child's Mind . . .

One More Is "More"

When young children think about *more than five*, they usually can imagine one more: six. Two or three more isn't as easy for them to conceptualize.

A Child's Mind . . .

Less and More

Young children seem much more interested and able to understand the idea of *more* than the idea of *less*. Find ways to incorporate the language of *less* in students' daily routines. Ask questions such as the following:

- "Are there less rainy days or sunny days this month?"
- "Are there less girls or less boys in this line?"
- "Which is less: the number of vowels or the number of consonants?"

Also consider displaying posters illustrating the concepts of *less* and *more*. These serve as important reference tools for second language learners as well. An example of this type of poster is provided as Reproducible 2, Greater Than, Less Than, and Equal To Reference Chart.

Explore

11. When students understand how to play the game, have them play it independently with their shoulder partner using a number path they have previously constructed on a sentence strip or a cash register paper tape. When I play this game with young students, I model writing my number on a scrap of paper and putting it in my pocket. That way, I keep the number a secret, but when it is correctly guessed, I show students the mystery number. It will be important to let each player have a turn to determine the mystery number as well as guess a partner's mystery number. To ensure both players get a turn, assign roles of who chooses a number and who guesses the number for the first round. Then, they switch roles. If some students need additional support, play another round (or more) with a small group to review the game.

Summarize

After students have had the opportunity to play the game with partners over a period of time (note that this game can be part of a math workshop learning station), gather them together for a whole-class discussion.

12. Ask key questions as well as the following:

    How did you discover the mystery numbers in your rounds of the game? Students might respond that they used the covered numbers to eliminate numbers that did not match the clues. Some may describe how they used numbers until they got to the mystery number. Some students may have tried guessing their own favorite numbers or the ages of their siblings. If students are playing the game using the extensions (see "Extend Their Learning!" at the end of the directions), they might respond with ideas about using the arrows to eliminate numbers that did not match the clues.

Which numbers were hard to guess? Why? When there were a lot of different guesses, students may think that these were the hard numbers. If they got lucky and guessed the mystery number quickly, they may believe that this mystery number was easy. Eventually students should think about trying a guess that matches all the given clues.

How were you able to guess the numbers on the open number path without any numbers? This question could be asked when playing the extensions (see "Extend Their Learning!" at the end of these directions). Accept all responses. Hopefully students will begin to see that they didn't need to count out each paper square to know where the ninth square was located but could use the two colors to find the 9 near 10 and/or counted on 4 more than 5.

13. Let students know that there are all different kinds of number paths and number lines. Some number paths and number lines are labeled with many numbers and some aren't. Show them an open number line labeled *0* and *10* and hint that they will be playing the *Mystery Number* game on this number line in the future.

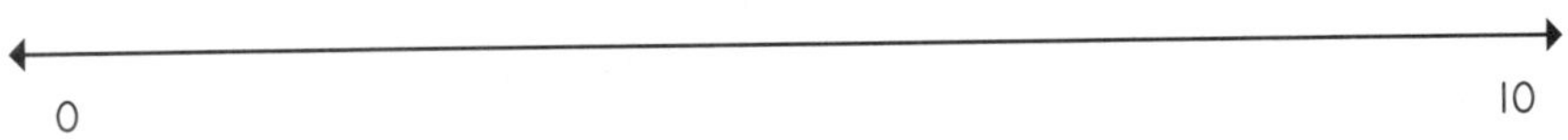

Open number line labeled *0* and *10*.

Extend Their Learning!

Kindergarten Version with a Larger Range of Numbers

When kindergartners are comfortable playing this game with a number path labeled *1–10*, extend the game to *1–20*, continuing to use sticky notes to cover the numbers eliminated by guesses.

Open Number Path Version

On another day, instead of using a labeled number path, have students use an open number path. This encourages them to count by ones, fives, or tens. Instead of guessing the cardinal number, students might be able to use the language of ordinal numbers. For example, a student might ask, "Is it the third square?" Instead of sticky notes, use clothespins marked with arrows (one clothespin with an arrow pointing to viewer's left to indicate "less than" and one arrow pointing to viewer's right for "more than") and move the clothespins back and forth on the number path according to the guesses. (Note that if you do not have clothespins, use two sticky notes with arrows drawn accordingly.)

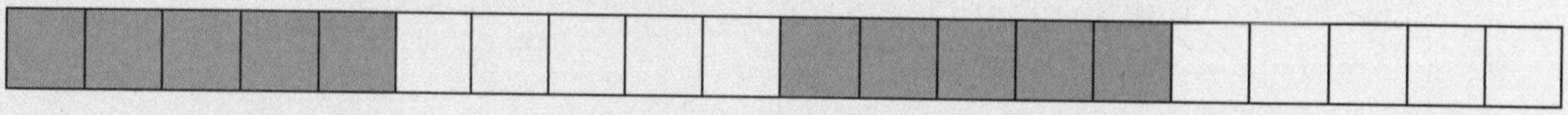

Open number path.

Example of Student Thinking

Teacher: "I'm thinking of a square that is somewhere on this number path."

Davit: "Is it number 10?"

Teacher: "The mystery number is more than ten. I'm going to put the clothespin on the tenth square here to show that the mystery number is more than ten. So it is somewhere along here." [Gesture to the remaining squares on the path.]

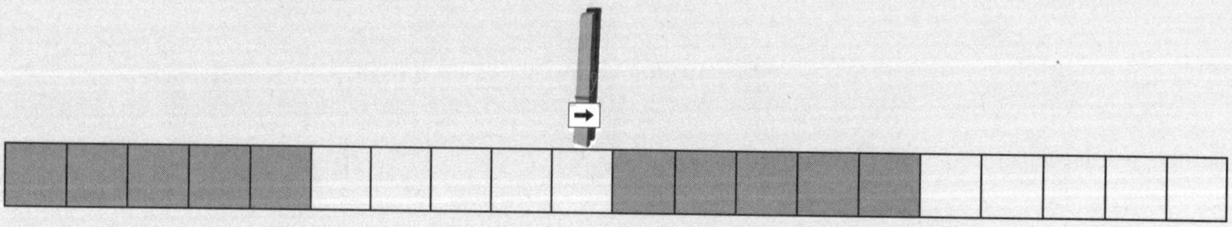

The mystery number is more than 10.

Dennis: "Is it number twenty?"

Teacher: "The mystery number is less than twenty. I'd like a volunteer to use this clothespin with an arrow indicating "less than" and show that the number is less than twenty on our number path."

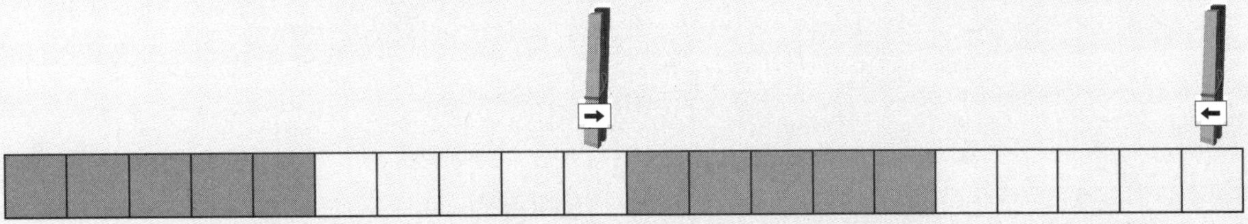

The mystery number is less than 20 and more than 10.

Sarah: "Is it twelve?"

Teacher: "The mystery number is greater than twelve. Which clothespin should we move to show that the number is more than twelve?"

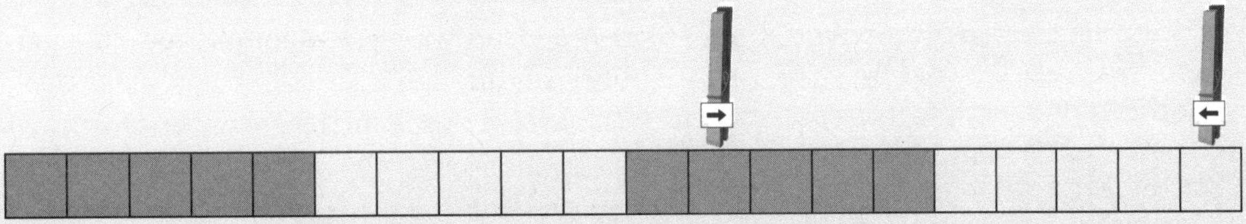

The mystery number is greater than 12 and less than 20.

Eva: "How about fifteen?"

Teacher: "The mystery number is less than fifteen."

Play continues until the mystery number has been identified. (Hint: It is less than 15 and more than 12.)

Math Matters!

Notice that the teacher is using *greater* and *more* in these examples. Children benefit from hearing these words in a context where they are trying to make sense of the meaning of the words.

Game 2 Mystery Number (Version 2)

Time

10–15 minutes

Materials

cash register tape or sentence strips (enough for an open number path up to 40)

red and yellow paper squares (approximately 3 inches by 3 inches), 20 of each color, organized in groups of five starting with red

Reproducible 2, Greater Than, Less Than, and Equal To Reference Chart (enlarged for classroom display)

clothespins with arrows on them, like this:

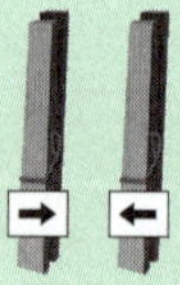

2 sticky notes with an arrow drawn facing left on one and facing right on the other, 1 set per pair of students

Overview

In this version of *Mystery Number*, instead of using a labeled number path like in *Mystery Number* (Version 1), students use an open number path to forty. (See Introducing Lesson 5, Using an Open Number Path without Numerals 1–60, for how to familiarize students with this open number path.) This encourages them to count by ones, fives, or tens. Students try to identify a mystery number on the number path through a series of guesses and clues. The holder of the mystery number gives clues using the words *less* or *more* and *less than* or *greater than*. The range of the guesses is eventually narrowed, supporting students in identifying the number. The examples in this lesson's directions are from a first-grade class; for other grades, increase or decrease the range of numbers on the number path.

Related Games

- G-1 Mystery Number (Version 1)
- G-3 Mystery Number (Version 3)
- G-11 Get to the Target (Version 1)
- G-12 Get to the Target (Version 2)
- G-13 Get to the Target (Version 3)

Key Questions

- What is your reason for your guess of ___?
- Why might your guess be a good guess?

Teaching Directions

Introduce

1. Prepare an open number path per the Materials list. Make sure the number path is large enough for everyone in the class to see it. Display it appropriately. For the purposes of these directions, an open number path to 40 is used. Set the two clothespins (per the "Materials" list) nearby.

Open number path to 40 (abbreviated here in order to fit on this page).

2. Gather students on the rug so that everyone has a good view of the number path. The students will need to be somewhat familiar with the idea of a numberless number path. They can count the squares and determine that there are forty squares, twenty are red and twenty are yellow. Share with students that they will be playing the game *Mystery Number*. Explain, "I am thinking of a number somewhere on this number path. Your job is to try to guess my number. I will give you clues each time you make a guess."

3. Choose a number. For the purposes of these directions, let's say that the number you choose is 22. This is the mystery number; do not share it with students. Consider writing the mystery number on a piece of paper and turning the paper over so students can't see it.

Teaching Tip

Choosing Numbers and Path or Line

When determining the range of numbers to use on a number path, be mindful of students' number knowledge. For instance, kindergarten students are likely more comfortable with a number path to 10 at the beginning of the year and could work with number paths to 20 later in the year. If you wish, you could begin this game with a smaller range of twenty paper squares (ten red and ten yellow) instead earlier in the year.

Teaching Tip

***Somewhere* on the Number Path**

Be careful of your language when describing the game. For example, if you say you are thinking of a number between 1 and 40, students might be confused about whether the number could be 1 or 40. For this reason, the directions are phrased intentionally to indicate to students that the number *is somewhere on the number path*. This also prompts students to think about the entire number path when making a guess.

Math Matters!

Symbolic Representations

In this game, the teacher provides the connection to the symbolic representation. Note, teachers often need to use the words *more* and *greater than* in the same sentence to remind students of the meaning of the symbol and the relationships.

Students are just beginning to make sense of the ideas of *less* and *more* with their guesses. Record clues using the greater than and less than symbols. Students in kindergarten and first grade need *many* meaningful opportunities to work with comparing quantities first before they are asked to make sense of (and write) their own symbolic representations.

4. Ask for a guess from a volunteer. Record the guess where everyone can see it, then respond to that guess with a clue. Note that instead of guessing the cardinal number, students might be able to use the language of ordinal numbers. For example, a student might ask, "Is it the tenth square?"

5. Show students the clothespins with the arrows and explain that they will be used in the game to show less and more. For the example, clip the clothespin with the arrow pointing right on the number path right above the tenth square and emphasize the meaning of the symbol by saying, "My number is greater than ten." Point to yourself when you say, "my number is," then to the → sign on the clothespin as you say, "greater than," and then finally, to the tenth square as your say, "ten"). Record the expression "______ > *10*" and say again, "My mystery number, which would fill in this blank, is greater than ten." Have students read the expression aloud.

Example of Student Thinking

Teacher: "I'm thinking of a square that is somewhere on this number path."

Maalai: "Is it the tenth square?"

Teacher: "The mystery number is more than ten. I'm going to put the clothespin on the tenth square here to show that the mystery number is more than ten. So it is somewhere along here." [Gesture to the remaining squares on the path.]

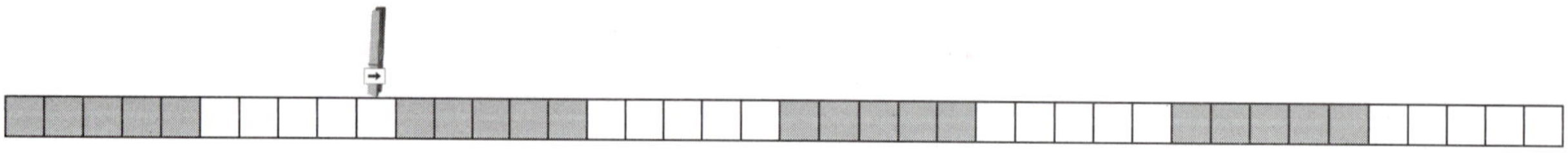

The mystery number is more than 10.

______ > *10*

6. Ask for another guess. Have students think about their guesses by asking key questions such as, "Why might that be a good guess?"
7. If the guess is thirty, for example, have a student volunteer clip the clothespin with the arrow pointing to the left above the thirtieth square to show that the mystery number is less than thirty. Use gestures to indicate the direction of *less than* as you say the words "less than." Write the following expression on the board: _____ < *30*. Then, say, using gestures, "My mystery number, which would fill in this blank, is less than thirty." Ask students to read the expression aloud.

Example of Student Thinking

Maxim: "Is it number thirty?"

Teacher: "Why might thirty be a good guess?"

Maxim: "Well, thirty is more than the last guess: ten. I know the mystery number is greater than ten, and I don't want to guess forty because it is too far."

Teacher: "The mystery number is less than thirty. I'd like a volunteer to use the clothespin with an arrow pointing to less and show that the number is less than thirty on our number path."

A Child's Mind . . .

Less and More

Young children seem much more interested and able to understand the idea of *more* than the idea of *less*. Find ways to incorporate the language of *less* in students' daily routines. Ask questions such as the following:

- "Are there less rainy days or sunny days this month?"
- "Are there less girls or less boys in this line?"
- "Which is less: the number of vowels or the number of consonants?"

Also consider displaying posters illustrating the concepts of less and more. These serve as important reference tools for second language learners as well. An example of this type of poster is provided as Reproducible 2, Greater Than, Less Than, and Equal To Reference Chart.

Using Clothespins with the Arrow Signs

Use of the clothespins with arrow signs provides a visual way of shrinking the range of possible numbers toward the mystery number. This helps students focus on the range of possible future guesses as well as shows how big jumps can reduce the range of guesses. This kind of thinking comes to students after they have had many opportunities to play this game. If you do not have clothespins, use two sticky notes, each with an arrow drawn on them accordingly.

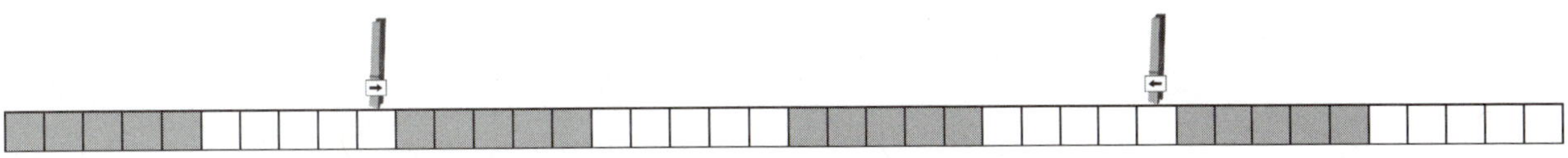

The mystery number is more than 10 and less than 30.

_____ > 10
_____ < 30

A Child's Mind . . .

Strategies for Efficient Guessing

Understanding which numbers to guess may not be readily apparent to students until they have played the game many times. One of the most efficient strategies for thinking about which number to guess in solving for the mystery number is to choose a number that is halfway across the range of remaining numbers.

Often, students guess a number that seems random but is important to them, such as their age or birthday. When size of the range is reduced dramatically and made apparent visually by covering numbers with sticky notes or moving the clothespin far across the range of possibilities, children begin to notice how quickly they can eliminate the greatest number of possibilities with one guess. This is an example of strategic and logical mathematical learning based on the thinking that is within the child's mind. It cannot be taught effectively to students by just explaining or showing. Until this idea becomes apparent to children, they may guess: "Is it one?" "Is it two?" "Is it three?" and so on.

8. Continue encouraging guesses, asking key questions, and recording and providing more clues based on the guesses. Eventually, the two clothespins will move closer together until the mystery number is identified.

Example of Student Thinking

Kyra: "Is it twenty?"

Teacher: "The mystery number is greater than twenty. We need to move the clothespin to show that the number is more than twenty."

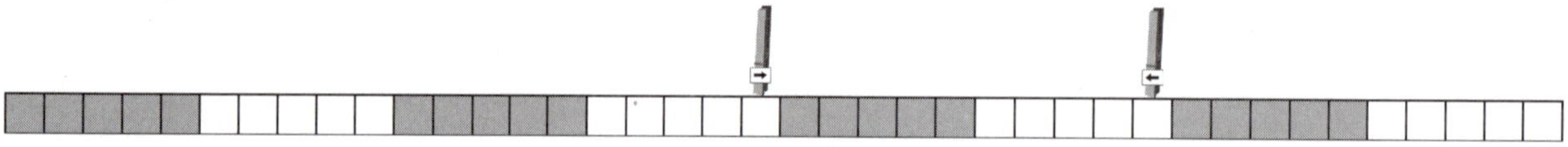

The mystery number is greater than 10, less than 30, and greater than 20.

______ > 10
______ < 30
______ > 20

Eva: "How about twenty-five?"

Teacher: "Why might twenty-five be a good guess?"

Eva: "Well, that twenty-five square is in between the two clothespins."

Teacher: "The mystery number is less than twenty-five."

Play continues until the mystery number has been identified.

Explore

9. When students understand how to play the game, have them it play independently with their shoulder partners using the "Greater Than, Less Than, and Equal To Reference Chart" (Reproducible 2). They can use two sticky notes with arrows on them to indicate number guesses and the range of remaining possible numbers to guess. When I play this game with young students, I model writing my number on a scrap of paper and putting it in my pocket. That way I keep the number a secret, but when it is correctly guessed, I show students the mystery number. It will be important to let each player have a turn to determine the mystery number as well as guess a partner's mystery number. To ensure both players get a turn, assign roles of who chooses a number and who guesses the number for the first round. Then, they switch roles. If some students need additional support, play another round (or more) with a small group to review the game.

Math Matters!

Two-Digit Numbers

Often, young children have difficulty reading two-digit numbers. They say ninety-one for nineteen or seventy-two for twenty-seven, for example. If necessary, read the number correctly for them, but do not stop at this time to explain the place-value concepts. Make a note of which students are confused about this and provide additional experiences at a later time to strengthen this skill. With more exposure to larger numbers and the patterns found in the place-value system, these students will begin to read two-digit numbers correctly. Display a hundreds chart on the wall that can serve as a reference for students who are still learning about place value and the counting sequence. (For more on the use of hundreds charts, see *It Makes Sense! Using the Hundreds Chart to Build Number Sense* by Melissa Conklin and Stephanie Sheffield [Math Solutions, 2012].)

Summarize

10. After students have had the opportunity to play the game with partners over a period of time (note that this game can be part of a math workshop learning station), gather them together for a whole-class discussion.

11. Ask key questions as well as the following:

    How did you discover the mystery numbers in your rounds of the game? Students should respond with ideas about using the arrows to eliminate numbers that did not match the clues. Some may describe how they used numbers until they got to the mystery number. Some students may have tried guessing their own favorite numbers or the ages of their siblings.

    Which numbers were hard to guess? Why? When there were a lot of different guesses, students may think that these were the hard numbers. If they got lucky and guessed the mystery number quickly, they may believe that number was easy. Eventually students should think about trying a guess that matches all the given clues.

    How were you able to guess the numbers when the number path didn't have any numbers? Accept all responses. Hopefully your students will begin to see that they didn't need to count out each paper square to know where the ninth square was located, for example, but could use the two colors to find the 9 near 10 and/or counted on 4 more than 5.

12. Tell them that there are all different kinds of number paths and number lines. Some have many numbers on them and some don't. Show them an open number line labeled *0* and *10* and hint that they will be playing the *Mystery Number* game on this number line in the future.

Extend Their Learning!

Justify Your Guess

Encourage students to articulate the reasoning behind their guesses by requiring them to justify their guesses. A justification must include one or two of the known clues in their guess. A possible sentence frame could be:

> My guess is _______ because it is more than _______ and less than _______ .

Game 3 Mystery Number (Version 3)

Time

10–15 minutes

Materials

open number line labeled *0* and *100*, created using a whiteboard or sheet of paper

2 sticky notes with an arrow drawn facing left on one and facing right on the other

Reproducible 2, Greater Than, Less Than, and Equal To Reference Chart (enlarged for classroom display)

In this version of *Mystery Number*, instead of using a number path as in *Mystery Number* (Version 1) and *Mystery Number* (Version 2), students use an open number line. They try to identify a mystery number on the number line through a series of guesses and clues. The holder of the mystery number gives clues using the words *less* or *more* and *less than* or *greater than*. The range of the guesses is eventually narrowed, supporting students in identifying the number. The examples in this lesson's directions are from a second-grade class; for other grades, increase or decrease the range of numbers on the number line. Students should have been introduced to the open number line in *Using an Open Number Line with Numerals 0 and 50* (I-10).

Related Games

- G-1 Mystery Number (Version 1)
- G-2 Mystery Number (Version 2)
- G-11 Get to the Target (Version 1)
- G-12 Get to the Target (Version 2)
- G-13 Get to the Target (Version 3)

Key Questions

- What is your reason for your guess of ___?
- Why might your guess be a good guess?

Teaching Directions

Introduce

1. Before beginning the lesson, draw an open number line where all students can see it. Label it *0* and *100*. Set the two sticky notes with arrows (per the "Materials" list) nearby.

Open number line labeled *0* and *100*.

2. Gather students on the rug so that everyone has a good view of the number line. Share with students that they will be playing the game *Mystery Number*. Explain, "I am thinking of a number somewhere on this number line. It might be here (point to one part of number line) or here (point to another part) or even here (point to another part). Your job is to try to guess my number. I will give you clues each time you make a guess."
3. Choose a number. For the purposes of these directions, let's say that the number you choose is 75. This is the mystery number; do not share it with students. Consider writing the mystery number on a piece of paper and turning the paper over so students can't see it.
4. Ask for a guess from a volunteer. Record the guess where everyone can see it, then respond to that guess with a clue.

Example of Student Thinking

Olivia: "Is it fifty?"

Teacher: "My number is greater than fifty. Your guess of fifty would be about here, in the middle." [Point to the middle of the number line.] Write *50* on the open number line.

Teaching Tip

Choosing Numbers

If you want the game to move quickly because of a limited amount of time, narrow the range of numbers. If you want students to work with larger double-digit numbers, use a number line that starts at 50.

***Somewhere* on the Number Line**

Be careful of your language when describing the game. If you say that you are thinking of a number between 1 and 10, students might be confused about whether the number could be 1 or 10. For this reason, the directions are phrased intentionally to indicate to students that the number *is somewhere on the number line*. This also prompts students to think about the entire number line when making a guess.

Math Matters!

Spacing

Approximate spacing on an open number line is important to model. You do not have an exact measurement of the distance between two numbers, but you do have a sense of one number as being greater than or less than another number because of its location on the number line. You could say something like, "Well, your guess of fifty would be about in the middle of this number line" or "Your guess of ninety would go about here, closer to one hundred, so I'm going to mark it here on our number line."

Math Matters!

Symbolic Representations

In this game, the teacher provides the connection to the symbolic representation. Note, teachers often need to use the words *more* and *greater than* in the same sentence to remind students of the meaning of the symbol and the relationships.

Students are just beginning to make sense of the ideas of *less* and *more* with their guesses. Record clues using the greater than and less than symbols. Students in kindergarten and first grade need *many* meaningful opportunities to work with comparing quantities first before they are asked to make sense of (and write) their own symbolic representations.

5. Show students the sticky notes with the arrows and explain that the sticky notes will be used in the game to show less and more. For the example, place the sticky note with the arrow pointing to the right where 50 has been recorded and emphasize the meaning of the symbol by saying, "My number is greater than fifty." Point to yourself when you say "my number is," then to the 50 and the sticky note as you say "greater than," and then finally to the middle of the open number line as you say "fifty." Record "______ > *50*" and say again, "My mystery number, which would fill in this blank, is greater than fifty."

Example of Teacher Recording

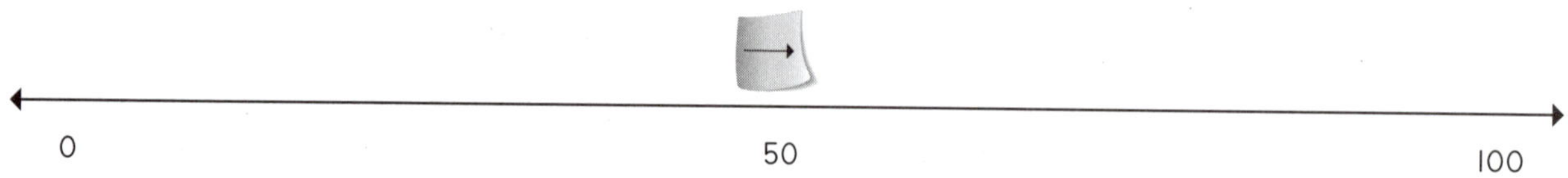

The mystery number is more than 50.

______ > 50

6. Ask for another guess. If the guess is greater than 75, such as 90, place the sticky note with the arrow pointing to the left on the number line closer to the 100, write *90* under the sticky note and also write the expression ____ < *90* under the first clue as in the following example. Read this expression by saying, "The mystery number is less than ninety." Add, "It is also greater than fifty." Ask students to do the same. Review the clues so far. "My number is greater than fifty and less than ninety."

Example of Teacher Recording

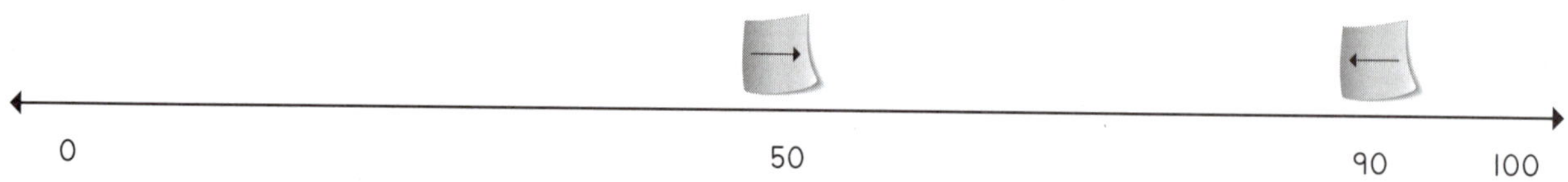

The mystery number is greater than 50 and less than 90.

______ > 50

______ < 90

7. Continue encouraging guesses, asking key questions, and recording and providing more clues based on the guesses. Eventually, the two sticky notes get moved closer together until the mystery number is identified. Sometimes the recording becomes crowded with number guesses near the end of the game. I usually use the entire whiteboard to draw the open number line all the way across in the beginning to allow for many guesses.

8. Eventually, you have a list of expressions that look something like the following example:

 ____ > 50

 ____ < 90

 ____ > 60

 ____ < 80

 ____ > 70

 ____ < 78

 ____ > 73

 ____ = 75!

Math Matters!

Two-Digit Numbers

Often, young children have difficulty reading two-digit numbers, saying ninety-one for nineteen or seventy-two for twenty-seven, for example. If necessary, read the number correctly for them, but do not stop at this time to explain the place-value concepts. Make a note of which students are confused about this and provide additional experiences at a later time to strengthen this skill. With more exposure to larger numbers and the patterns found in the place-value system, these students begin to read two-digit numbers correctly. Display a hundreds chart on the wall that can serve as a reference for students who are still learning about place value and the counting sequence. (For more on the use of hundreds charts, see *It Makes Sense!: Using the Hundreds Chart to Build Number Sense* by Melissa Conklin and Stephanie Sheffield [Math Solutions, 2012].)

Math Matters!

The Equal Sign

The expression "____ = 75" may look new to students. They may have only been familiar with seeing the equal sign used with two addends to the left of it and the sum to the right. In our example, the expression can be read as "The mystery number is equal to seventy-five," or "The mystery number is seventy-five." This helps children begin to understand the meaning of the equal sign as an expression showing a relationship.

A Child's Mind . . .

Less and More

Young children seem much more interested and able to understand the idea of *more* than the idea of *less*. Find ways to incorporate the language of *less* in students' daily routines. Ask questions such as the following:

- "Are there less rainy days or sunny days this month?"
- "Are there less girls or less boys in this line?"
- "Which is less: the number of vowels or the number of consonants?"

Also consider displaying posters that illustrate the concepts of *less* and *more*. They serve as important reference tools for second language learners as well. An example of this type of poster is provided as Reproducible 2, Greater Than, Less Than, and Equal To Reference Chart.

Teaching Tip

Groups of Four

Some teachers like to have students work in groups of four students. Two students are the keepers of the mystery number, and two students are the guessers. Pairs can support and talk with one another about how to record the guesses on the open number line and get help from the guessers about where to approximately locate the guess and the sticky note arrows. Partners need to agree on the guesses they offer, making sure that the guess is a number that could be described by the clues that have been given.

Explore

9. When students understand how to play the game, have them play independently with their shoulder partners. They can draw their open number line on a blank piece of paper, using a range of 1–100 (or 1–50 if you want them to work with smaller numbers at the beginning). They can draw arrows or use sticky notes to indicate eliminated numbers. When I play this game with young students, I model writing my mystery number on a scrap of paper and putting it in my pocket. That way I keep the number a secret, but when it is correctly guessed, I show students the mystery number. It will be important to let each player have a turn to determine the mystery number as well as guess a partner's mystery number. To ensure both players get a turn, assign roles of who chooses a number and who guesses the number for the first round. Then, they switch roles.
10. If some students seem confused, play another round (or more) with a small group to review the game.

Summarize

11. After students have had the opportunity to play the game with partners over a period of time (note that this game can be part of a math workshop learning station), gather them together for a whole-class discussion. Have students bring their number line drawings and share them with the class, describing the action of the game from the first tentative guesses.

A Child's Mind . . .

Strategies for Efficient Guessing

Understanding which numbers to guess may not be readily apparent to students until they have played the game many times. One of the most efficient strategies for solving for the mystery number is to choose a number that is halfway across the range of remaining numbers.

Often, students guess a number that seems random but is important to them, such as their age or birthday. When the range is reduced dramatically and made apparent visually by the arrow-marked sticky notes, children begin to notice how quickly they can eliminate the greatest number of possibilities with one guess. This is an example of strategic and logical mathematical learning based on the thinking that is within the child's mind. It cannot be taught effectively to the student by just explaining or showing.

Assessment Opportunities

Show students the results of one game and ask them to re-create the number line from the given information.

For example:

___ > 10

___ > 20

___ < 50

___ > 30

___ < 34

___ > 32

What's the mystery number?

Game 4

Race to 50

Time

30 minutes

Materials

Curious George Learns to Count from 1 to 100 by H. A. Rey (or another favorite counting book)

open number lines labeled *0* and *50*, 1 marked *Red Team* and 1 marked *Blue Team*, created on a whiteboard or using a large sheet of paper, 1 set per each red team versus blue team

red marker, 1 per each red team

blue marker, 1 per each blue team

Race to 50 Action Cards (Reproducible 13), 1 set per each red team versus blue team (made from 1 copy of the reproducible)

Race to 50 Question Cards (Reproducible 14), 1 set per each red team versus blue team (made from 1 copy of the reproducible)

2 paper lunch sacks, 1 labeled *Action Cards* and 1 labeled *Question Cards*, 1 set per each red team versus blue team (to hold the cards)

Overview

This fast-moving game provides engaging opportunities for students to practice adding and subtracting, tracking their mental calculations, and building fluency. Teams alternate taking turns, first drawing an action card to determine their move on an open number line (adding or subtracting one, five, ten, or twenty). The team then draws a question card, which prompts them to think more about the number relationships and, again, to practice adding or subtracting.

Because this game is played on an open number line, students have a chance to show jumps in ways that match their level of understanding. For example, if the action card is +10, some students may choose to move by ten jumps of ones whereas others will move one jump of ten.

Race to 50 is most appropriate for kindergarteners and first graders. When students are comfortable with the game, extend the number line to 100 and play *Race to 100*. For second graders, play *Race from 100 to 200* using a 100-to-200 number line. See the "Extend the Learning!" section at the end of the game directions for insights on these variations.

Related Lessons

- L-4 Jumping by Ones and Tens
- L-5 Jumping by Ones, Fives, Tens, and Twenty

Key Questions

- How did you know how far to jump on the number line?
- How did you solve what was asked of you on the question card?
- Which card would you like to draw next? Why?
- Which card is not a very helpful card to draw? Why?

Materials for Extension

Race to 100 Action Cards (Reproducible 15), 1 set per each red team versus blue team (made from 1 copy of the reproducible)

Race to 100 Question Cards (Reproducible 16), 1 set per each red team versus blue team (made from 1 copy of the reproducible)

Time Savers

Preparing the Number Lines

Instead of making consumable copies of the number lines, laminate the copies or place them in plastic sleeves and provide dry-erase markers (red and blue) during game play. You may also have students draw their own number lines; in this case, distribute whiteboards and black dry-erase markers.

Preparing the Cards

For the purpose of this game, a set of cards is made with one copy of the reproducible. Copy the action cards and question cards on different-color paper so they can be sorted easily. Shuffle the cards and place them into paper lunch sacks (or in two piles—an action cards pile and a question cards pile).

Teaching Directions

Introduce

1. Prepare the demonstration open number lines per the "Materials" list. Make sure to label one *Red Team* and one *Blue Team*. Display them on the wall at student eye level.

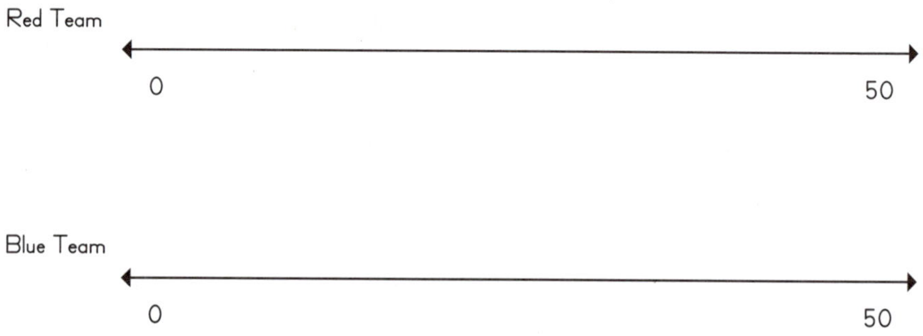

The two open number lines to use when playing *Race to 50*.

2. Prepare a demonstration set of both the action cards and the questions cards per the "Materials" list. To model the game, place each set in a pile. (After a few rounds of modeling the game, the cards can then be placed in paper lunch sacks.) Also, have a red marker and a blue marker on hand.

3. Gather students so that everyone has a good view of the number lines and card piles. Seat students close together to make listening to each other easier.

4. If you have the book *Curious George Learns to Count from 1 to 100* by H. A. Rey, set the stage for counting larger numbers by reading the pages that deal with counting up to fifty. If you do not have this book, choose one from your collection that supports counting.

Literature Connection

Curious George Learns to Count from 1 to 100

In the book *Curious George Learns to Count from 1 to 100,* George finds counting from one to ten easy, but can he count all the way to one hundred? Filled with George's delightful monkey mischief, this book is an especially friendly way to review the potential "stumbling block" numbers that come after twenty-nine, thirty-nine, and forty-nine.

5. Draw students' attention to the number lines and introduce the game. Say, "This game is played in teams. There will be two teams: a red team and a blue team." Divide the students into two teams and indicate that one team is the red team and the other is the blue team.

6. Continue, "The objective of the game is to get to fifty or get to more than fifty on your team's number line before the opposing team does the same on their number line. Teams take turns drawing action cards (point to the pile) to determine their move on the number line, beginning at zero."

7. Demonstrate a move first by choosing a student on the red team to draw an action card. Use questions to guide the class in making the corresponding move on the number line. For example, if Alicia on the red team draws a +20 card, say, "Alicia drew a plus-twenty card, meaning she can move to twenty. How can Alicia get to twenty? Who has another way? How does the red team want to record its jump?" If the red team decides to take two jumps of ten for the +20 card, use the red marker to record the jumps on the corresponding number line:

Teaching Tip

Games with Teams

When introducing team games, roughly divide the class in half (right half of the class versus the left half of the class). They can move their chairs close to one another or huddle together on the rug to discuss their moves and answers to the game cards. Let students know you will call on one representative each turn. They will be able to take turns being the spokesperson for the group. When teams get a chance to discuss their moves and responses, they often make better decisions and learn strategies from one another.

A Child's Mind . . .

Making Moves on a Number Line

Eventually, students will keep track of their own jumps; but, while modeling the game, it's best that the teacher do the recording. Note that, at first, students may need to make jumps on the number line in increments of ones, fives, and/or tens. After several turns or games, students may be able to jump in multiples of tens. Encourage larger jumps. Also, always encourage students to explain how they know they landed on and recorded the correct number.

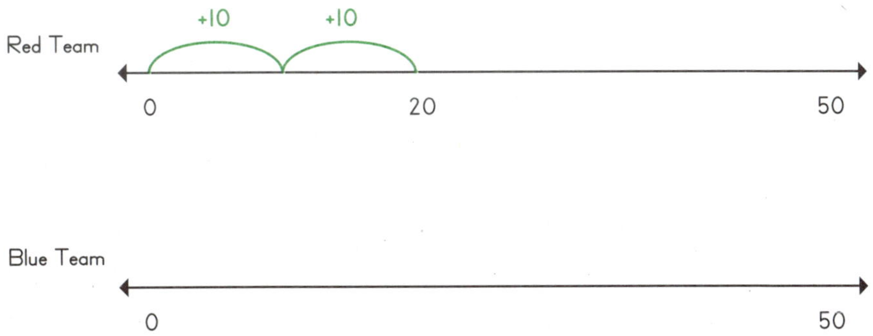

The red team's move after drawing the action card +20.

Teaching Tip

Spacing on an Open Number Line

It's important to pay attention to and demonstrate approximate spacing on an open number line. You do not have an exact measurement of the distance between 0 and 50, but students should have a sense of how the numbers increase by their location on the number line. You could say something like, "Well, your jump of twenty would be about here on our number line, which is not quite halfway to fifty, so we can mark it here. We're not worrying about being exact; we just want to show the jump of twenty."

Differentiating Your Instruction

Supporting English Language Learners

The question cards may need to be omitted or modified in this game depending on students' ability to read. You may choose to have students work with just one question card and task each team with answering it every time the teams complete an action card.

8. Model a few rounds using just the action cards (teams take turns drawing a card, deciding their move, and recording their thinking on the corresponding number line). If a team draws a –10 card that would move them back before zero, have them place the card at the bottom of the pile and draw again.

9. Now introduce the use of the question cards. Explain that after each team draws an action card and records its move on the number line, the team must then draw a question card. Emphasize that a different team member should draw the question card than the action card, so as many team members as possible are participating in drawing the cards. Ask a student on the red team to draw a card and read the question to the team.

10. Have students on both teams turn and talk about how to respond to the question, then ask the red team members to share their answer. If the question card drawn is *What is 10* more *than what you have right now?*, a red team member might respond, "Ten more than twenty is thirty." It can be helpful to ask students to explain how they know that "ten more than twenty is thirty." Some students will count on by ones from twenty while others will point to a hundreds chart and show the multiples of tens in that column.

11. Now have the blue team demonstrate a full turn using both a new action card and a new question card. Use key questions to support students in thinking about how the game works.

Examples of Student Thinking (from the blue team)

Cyrus [after drawing an action card and conferring with his team]: "I got a plus-ten card, so I'm going to jump to ten."

Teacher [after recording Cyrus's thinking]: "Does this show what you're thinking?"

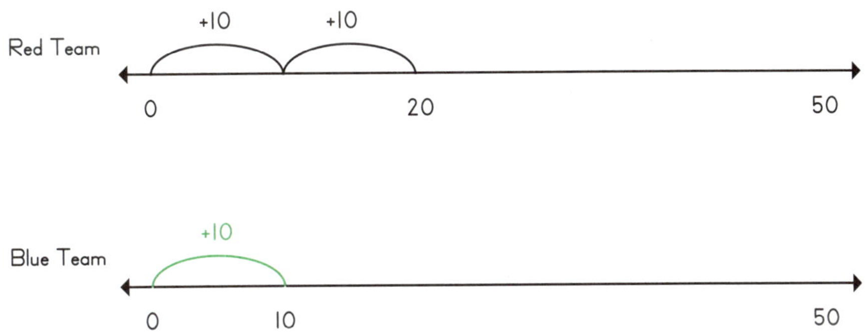

The blue team's move after drawing the action card +10.

Cyrus: "Yes!"

Seiko [after drawing a question card and conferring with his team]: "Our question card says: What is five less than what you have right now? Five less would be five 'cause five and five are ten."

Teacher [in an effort to further engage thinking]: "Which card would you like to draw next? Why?"

Erin: "We want a plus twenty so we can be ahead of the red team."

Teacher: "Which card is not a very helpful card to draw? Why?"

Marcos: "Well, if we got a plus one, we wouldn't move very far."

12. Now have the red team demonstrate a full turn using both the action cards and the questions cards. Use key questions to support students in thinking about how the game works.

Teaching Tip

The –10 action card presents a recording challenge. To keep the recording from becoming too messy or confusing, show this jump going below the number line, possibly using dotted lines.

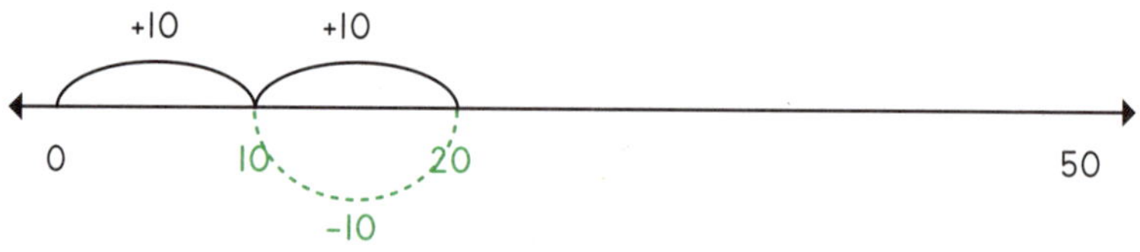

The red team's next move after drawing the action card -10.

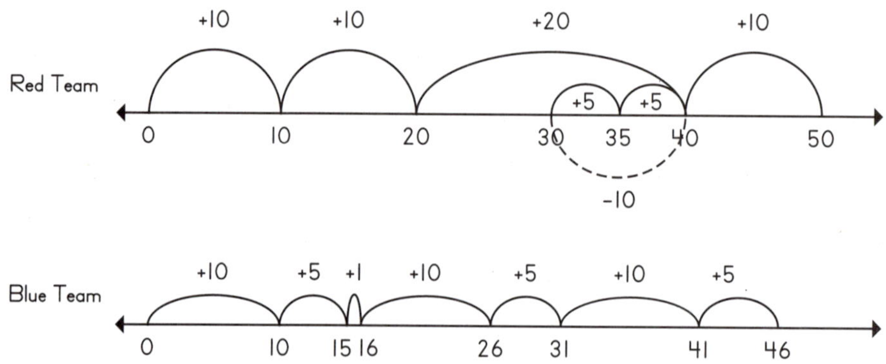

Both teams' number lines after a completed game.

13. Summarize the game by asking some more of the key questions.

Examples of Student Thinking

Teacher: "How did you solve what was asked of you on the question card? You were at sixteen and the card asked you to tell what is ten more?"

Raj: "We counted up to the number by jumping ten and got to twenty-six."

Andrei: "We know that ten and ten are twenty so there's six more."

Eshwari: "You just look at the number and then see that you need ten more right here [pointing to the tens place]."

Explore

14. After students are comfortable with the game procedures, have them play the game in small groups. Pair them into teams of two and join a red team with each blue team. Make sure each group has the number lines, red and blue markers, action cards, and question cards per the "Materials" list.
15. Encourage students to work together. Say, "When you play this game in teams of two, take turns with your teammate in choosing the action card and the question card, so you both get a chance to do both tasks. Also, involve your partner by listening to their strategy before you record the move on your team's number line. After you've recorded a move, check each other's work. Does your recording match the action card?"
16. Help groups decide which team goes first. One way to do this is to have groups shake up their paper sacks of action cards and then have each team draw a card. The team that chooses the card with the greatest number on it goes first. (If they get the −10 card they draw again.)
17. Circulate as students play, asking key questions to guide their thinking (see the Key Questions list on page 193).

Summarize

18. After students have had the opportunity to play the game independently over a period of time (note that this game can be part of a math workshop learning station), gather them together for a whole-class discussion.

Differentiating Your Instruction

Pairing Students
For this game, pair students with someone of like ability. Pairing students with similar mathematical skills may allow for more differentiation options and will afford optimum access to the mathematics involved in playing the game.

A Child's Mind . . .

Looking for Patterns
After continued play, some students will be ready to work with numbers beyond fifty on the number line whereas others will still be building their comfort with numbers to fifty. Encourage students who are still developing number sense for numbers less than fifty to look for patterns in the counting numbers. It's often helpful to post a hundreds chart (one through one hundred) in the room for reference. (A reproducible hundreds chart is available in *It Makes Sense!: Using the Hundreds Chart to Build Number Sense* [Math Solutions, 2012], also in this series.)

Extend Their Learning!

Race to 100

When students are comfortable and familiar with the game *Race to 50,* extend the number line to one hundred and play *Race to 100.*

The two open number lines to use to play *Race to 100.*

The game directions of this extension are the same as *Race to 50*; however, students use different action cards (Reproducible 15) and question cards (Reproducible 16). The addition of the +20, +30, +40, and –20 action cards make the game move more quickly to one hundred than the action cards used to play *Race to 50.*

The following are examples of student thinking for this version of the game when the question card, *How far from 100 are you?* is drawn.

Examples of Student Thinking

Arturo: "From twenty to one hundred is ten plus ten plus ten plus fifty, or eighty jumps."

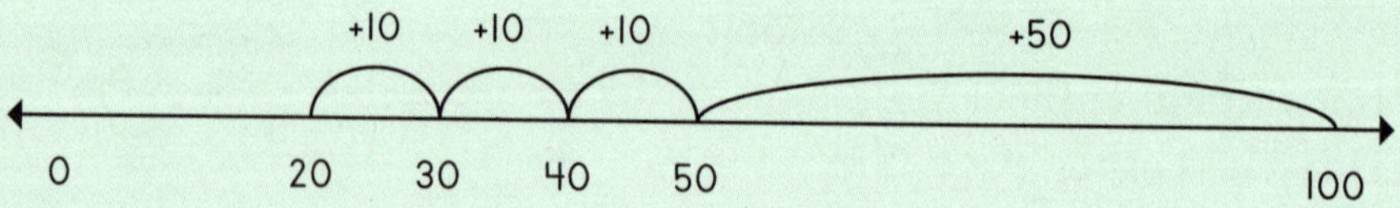

An example of Arturo's thinking.

Sasha: "From fifty-six to one hundred is four plus forty, or forty-four jumps."

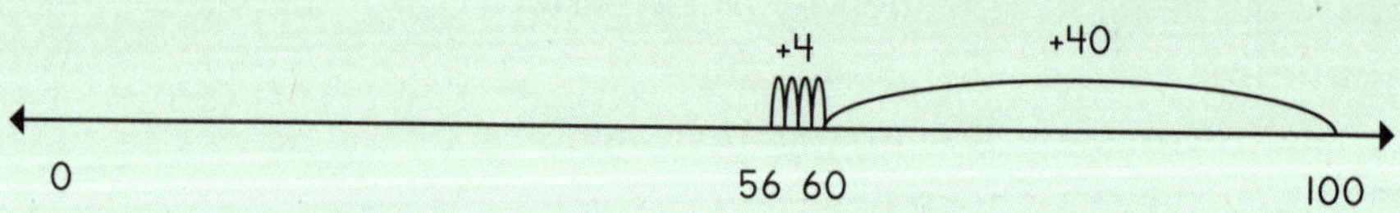

An example of Sasha's thinking.

The following is an example of how the summarizing part of the *Race to 100* version of the game might unfold when the first team has already moved to 5 on the open number line in their first turn.

1. Display the action card +40 and ask students to think quietly for a few moments about the different ways they could jump forty spaces from the 5 on the number line. Then have students turn and talk to a shoulder partner so they can share their strategies.
2. Call on a few students to share their thinking. Have the students come to the front of the class, point to the number line, and show what they mean.
3. Record below the number line the equation that each student demonstrates.

Examples of Student Thinking

Amal: "Ten plus ten plus ten plus ten is forty, so I jumped four tens from five and got to forty-five."

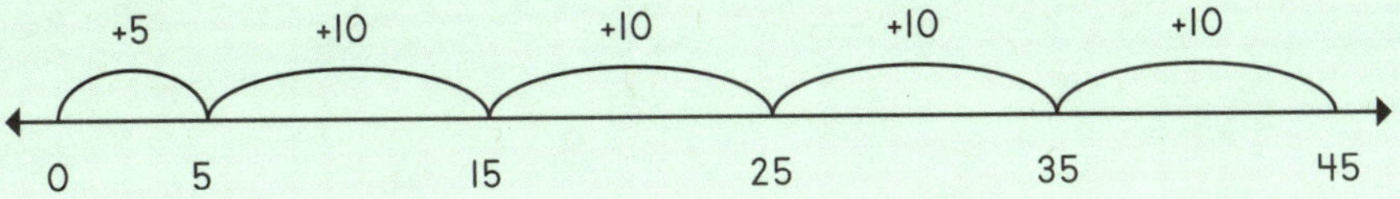

Teacher [after recording Amal's thinking]: "Does this show your thinking?"

10 + 10 + 10 + 10 = 40

5 + (10 + 10 + 10 + 10) = 45

Amal: "Yes."

Ramona: "I just jumped by fives because I know how to count by fives. I did it eight times because I used eight fingers to count to forty. I ended up at forty-five."

Teacher [after recording Ramona's thinking]: "Does this show your thinking?"

5, 10, 15, 20, 25, 30, 35, 40

5 + 5 + 5 + 5 + 5 + 5 + 5 + 5 = 40

5 + (5 + 5 + 5 + 5 + 5 + 5 + 5 + 5) = 45

Ramona: "Yes."

Teaching Tip

Introducing Parentheses into Student Thinking Recordings

When recording what a student says, consider using parentheses when appropriate. Explain to students that this is the way mathematicians organize their thinking when a lot is going on. As you draw the parentheses, say something like, "This tells us that we have this group of fives from when you counted by fives, and this five was your number where you started." Don't necessarily expect students to start using parentheses (younger children are just learning to write numbers and relational symbols) but do expose them to this way of representing their thinking. If using parentheses is uncomfortable for you, simply draw a thought bubble to record the student's thinking. Either way, the goal remains the same: putting student thinking down in writing so they can connect to the mathematical symbols what they think and say.

A Child's Mind . . .

Decomposing Numbers

In both the examples from the class discussion, students saw a larger number, such as 40, and decided to decompose it into more manageable parts. It is not efficient to change it into forty little parts; however, some students will continue to do this and count one-by-one if they are not comfortable with making larger chunks of the whole. When children think flexibly about a number like 20, for example, they can decompose it into two groups of ten, four groups of five, five groups of four, or ten groups of two, depending on the context in which they are working. (If they are thinking about chopsticks, then ten groups of two makes sense. If they are thinking about nickels, then four groups of five makes sense. A context for five groups of four might involve thinking about the number of wheels on five cars.)

We know that children are making sense of the numbers they are working with if they can comfortably decompose a number like 48 into four groups of ten and eight ones, two groups of twenty-four, or five groups of ten with two missing.

Race from 100 to 200

Race from 100 to 200 is especially appropriate for second graders and uses number lines labeled with *100* and *200*.

The two open number lines to use when playing *Race from 100 to 200*.

The game directions are the same as *Race to 50*; however, students use the same action cards and question cards as used in the extension *Race to 100*. On the number lines, you can go higher with the range of numbers (200 to 300 or 300 to 400) if students want to work with even larger numbers; sometimes students think larger numbers are more interesting!

Math Matters!

Two Common Strategies That Students Use

In making the jumps-of-ten strategy, one number is kept whole and jumps of ten are added to it. For example, 38 + 34 = 38 + 10 + 10 + 10 + 4 (the 4 is decomposed to 2 + 2 so the jump from 68 can go to 70, which is a landmark number. Then it is easier to jump two more to 72).

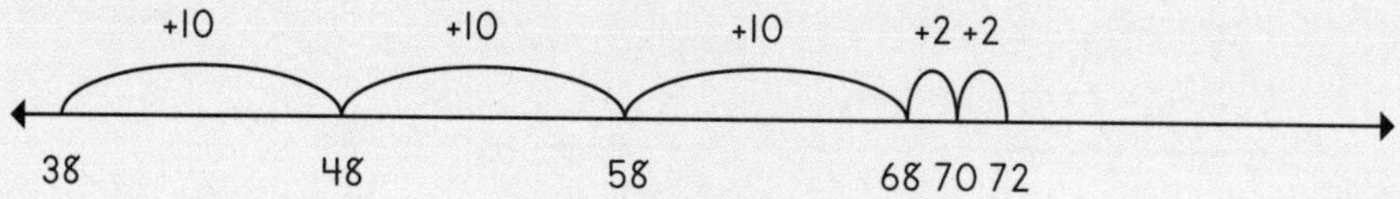

The jumps-of-ten strategy on an open number line.

With larger amounts, hundreds are added first, in the same way.

In the friendly-number strategy (sometimes called *compensation*), the units are added first to reach a friendly number. In an example like 48 + 56, students are likely to notice how close forty-eight is to fifty and make use of this friendly number, changing the forty-eight to fifty by giving two from the fifty-six to the forty-eight, making the problem 50 + 54, which is easier to solve.

In the same way, sixty-eight is close to seventy, which is a friendly number. So, the six in twenty-six is decomposed to 2 + 4 and then seventy becomes a friendly number, where 68 + 26 can be shown as:

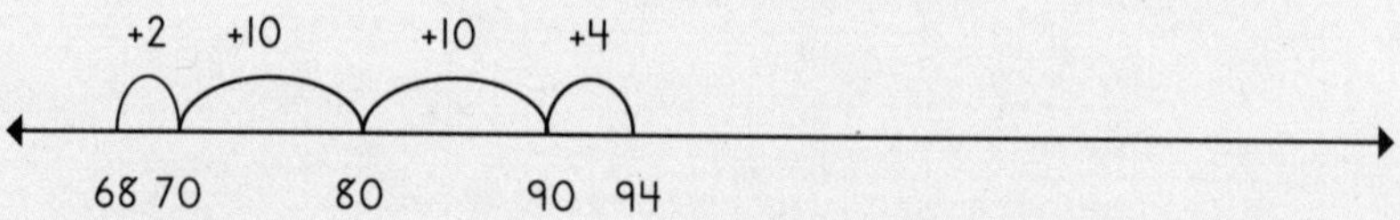

The friendly-number strategy on an open number line.

I have found that it is useful to make personal note of these strategies when students begin to use them. For example, when Dennis went to a friendly number of seventy from sixty-eight, we called this Dennis's friendly-number strategy. When Eva counted ten more from thirty-eight to forty-eight, we called this Eva's jumps-of-ten strategy. Students began to refer to the strategies by name, eventually telling the class that they were now using their own friendly-number strategy. I think having words to explain the ideas they were using helped them communicate their thinking to others and also made them proud of their own learning.

The Game of Pig on a Number Line

Game 5

Overview

In this game, the objective is to be the first player or team to reach 100 on a number line. Players or teams alternate turns, rolling two dice to determine how far they advance on their individual number lines. They may roll the dice as many times as they want during their turn, building on their last move each time. However, there may be penalties if players get too greedy (in other words, become a pig)! If a player rolls a 1, the player's turn ends and the moves for that round are taken away. In addition, if a player rolls two 1s at the same time, the player's turn ends and the player must clear their number line of all moves, returning to 0. These additional constraints make this game an active and engaging gamble because it's up to each player to decide when to quit and when to continue on in hopes of getting closer to 100. The game is introduced to the whole class, and then can be played in small groups with two or more players over time. It is most appropriate for first and second graders. Some kindergarteners might be able to play this game using a number line to 50 later in the year; however, thinking about adding the numbers on the dice *and* the likelihood of some combinations is much more challenging than simply racing to 50 or 100 as in Game 4, *Race to 50.*

Time

30–45 minutes

Materials

open number line labeled *0* and *100*, 1 per student

dice (labeled *1–6*), 2 for each group of 2 or more students

markers in different colors, 2 or 3 per group

felt placemats for dice rolls (optional)

Materials for Extension

pencil, 1 per pair of students

The Game of Pig on a Number Line Recording Sheet (Reproducible 17), 1 enlarged for classroom use and 1 per pair of students

Teaching Tip

Quiet Dice

Rolling dice can create a lot of noise. To lessen the noise, use foam dice or pad students' workspaces with foam or fabric placemats.

Time Savers

Preparing the Number Line

Instead of making consumable copies of the number line, laminate the copies or place them in plastic sleeves and provide dry-erase markers in different colors during game play. You may also have students draw their own number line; in this case, distribute whiteboards and black (and two other colors) dry-erase markers or provide blank paper for recordings.

Related Game

- G-4 Race to 50

Key Questions

- What are your moves on the number line so far? Why did you make those moves?
- Do you think you should take another turn? Why or why not?
- How far away are you from 100? How do you know that?

Teaching Directions

Introduce

1. Prepare a demonstration open number line per the "Materials" list. Display it on a white-board or on the wall at student eye level.

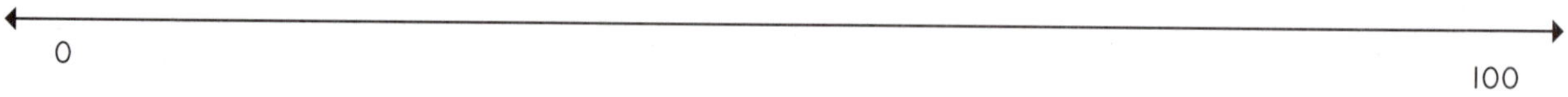

The open number line for use with *The Game of Pig on a Number Line*.

Technology Tip

Using an Interactive Whiteboard

If you choose to use an interactive whiteboard, draw the open number line on the screen with the interactive pen. In addition, you may choose to use the interactive die, so that all students can see the number rolled.

2. Gather students on the rug so that everyone has a good view of the number line. Seat them close together to make listening to each other easier.
3. Draw students' attention to the number line and introduce the game. Say, "We are going to play a game called *Pig*, and in this game, you want to be the first to get to one hundred or more on the number line. Today we are going to play as a whole class and see if we can get to one hundred or more."
4. Explain that the roll of two dice determines each player's move on the number line.
5. Introduce the game constraints. Say, "When it is your turn, you may roll the two dice as many times as you want during your turn, building on your last number line move

each time. However, there may be penalties! If you roll a one, your turn ends and the moves for that round are taken away. Also, if you roll two ones at the same time, your turn ends and you must clear your number line of all moves, returning to zero."

6. Write the constraints where everyone can see them:

Game Constraints
If you roll one 1, you must take away any moves you've made during that turn.
If you roll two 1s, you must remove all your moves and start from 0 again.

7. Check for understanding. Ask, "What happens if you roll a five and a one on your first roll?" (Because you rolled a 1, your turn ends, and you don't get any moves for that round.) "What happens if you roll two ones?" (You must remove all your moves and start from the beginning of your number line—0—again.)

8. Demonstrate the game as a class. Say, "We are going to play a game with the whole class to see if we can get to one hundred or beyond. We don't need to land exactly on one hundred." Randomly choose a student to go first for the class. Ask the student to roll the two dice. If, for example, the student rolls a 4 and a 3, have the class add the two numbers. The sum, 7, is how far the student can move on the class number line. Select a marker color and record the jump to 7 on a number line with that marker.

Teaching Tip

Managing the Use of Dice

When rolling dice, sometimes students are overly enthusiastic and throw the dice high in the air or across the room. Set classroom norms for using dice. If using placemats, caution students that they will lose their turn if they roll the dice off their placemat. Use similar consequences if the dice are not handled responsibly. One way to keep dice safely in play is to put the dice in a small, empty, clear water bottle. This makes an exciting rattling sound, but keeps the dice contained in a small space. Whatever you choose to do, being proactive in curtailing the misuse of dice makes games much safer and more manageable.

The number line with a first move shown: a roll of 4 and a roll of 3, for a total move of 7.

Decide, for demonstration purposes, that the student should roll again. This time, for example, the student rolls 5 and 6. This sum is 11. Add 11 to 7 for a total of 18. How might this be shown on the number line? Ask students to volunteer their thinking and record this move on the number line. (Note that in this example, the student added 10 and 1 instead of adding 11 to the first sum. Another student might decompose the 5 on the first die to 3 and 2, add 3 to 7 to get ten and then add the remaining 2 and 6 to get to 18.)

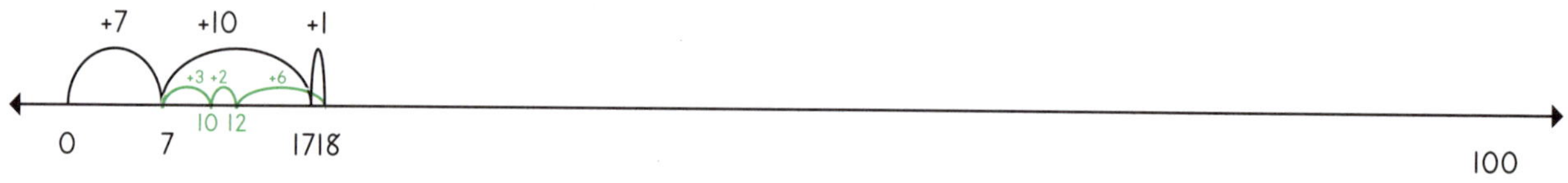

The number line with a second move shown: a roll of 6 and 5 for a total move of 11.

9. Ask another student to roll for the class. Let's say this student rolls a 5 and a 5. Have students add the numbers for a sum of 10. Ask the student, "How will you make this move on the number line?" If the student responds that they would "jump ten," using a different-color marker, add ten more to eighteen, show the jump on the number line, check with the student to ensure you understood their thinking, and ask the student if they want to continue rolling the dice.

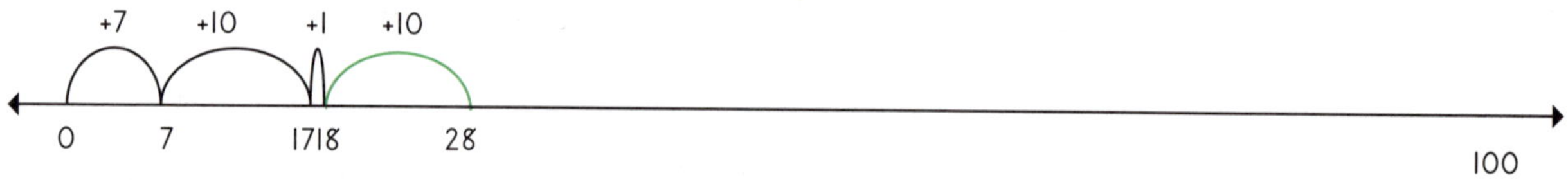

The number line with a third move shown: a roll of 5 and 5, for a total move of 10.

10. In our example, the student decides to roll again and gets a 2 and a 5, for a sum of 7. To make this move on the number line, the student explains that they add the 2 to 28 to get to 30, and then add the 5 to get to 35. Record the move.

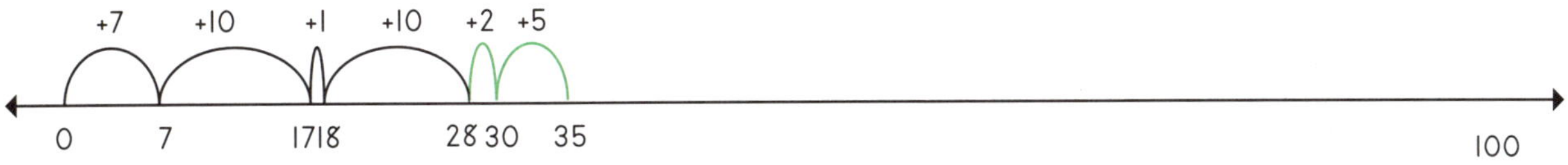

The number line now with a fourth move shown: a roll of 2 and 5, for a total move of 7.

11. This is a good time to ask key questions: Do you think you should take another turn? Why or why not?

Example of Student Thinking

Karolina: "Two people have rolled four times and we're already at thirty-five without getting a one, so I just think I don't want to be a pig."

Teacher: "How many more do you need to get to one hundred?"

Karolina: "It's fifteen to get to fifty and fifty more, so it's sixty-five to get to one hundred."

12. Ask another student to roll the dice. This time, in our example, let's say the student rolls a 1 and a 6. Because a 1 is rolled, the student's turn is over and the class stays at 35 on the number line. (Nothing more is recorded on the number line for this turn.)

13. Continue modeling the game, changing the color of the marker you use for each new player. If two 1s are rolled, erase all the number line moves and start over. If a 1 is rolled in the middle of a round, only the numbers for that round are erased.

Explore

14. After the game has been played with the whole class as one team, divide the class into two teams and begin to play a full, competitive game. Record the moves using a different number line for each team, until one of the teams gets to 100 or beyond. Because the directions may be confusing at first, the two-team class game might be better played over an extended period of several days, for

> **Teaching Tip**
>
> **Recording Students' Thinking**
>
> When recording moves on a number line during the demonstration part of the game, use one color for the first player in the first round, and another colored marker for the second round, so players can see where one round ends and the second one begins. Record the jumps and the numbers on the number line as students explain what they want to do with their roll. Sometimes students decompose the numbers; sometimes they just add the sum of the two dice; sometimes they count on from their number the results of one die at a time. It is important that everyone listen carefully to each student's explanation and that you record the addition using the same techniques the students choose so they see a representation of their own thinking.

about ten minutes at a time. Continue to use the key questions listed earlier in this game.

15. Encourage students to tell their team members their reasons for whether they should roll again.

16. Students may continue to play the game as a class or play it independently in small groups of two or more players. When students play in small groups, they form two teams and keep track of their own team scores on their own recording sheets or erasable open number lines. Read more in the "Extend Their Learning!" section at the end of the game.

Summarize

17. After students have had the opportunity to play the game independently over a period of time (note that this game can be part of a math workshop learning station), gather them together for a whole-class discussion about the greedy part of *The Game of Pig.* Ask the following questions and let a few students display their recording sheets to the whole class:

    - Did anyone get almost to 100 and then roll two ones?
    - Who had a long run of jumps without rolling any ones?
    - Were any of the games really close?
    - What makes this a fun or interesting game?

Extend Their Learning!

Recording the Rolls

After students have played the game several times over the course of at least a few days and they are comfortable with the procedures, they may begin to notice that the chance of rolling a 1 (or two 1s) is less frequent than other rolls. Now would be a good time to spend some time outside of the game to collect evidence to determine why some combinations of two dice are not as common as other combinations. Ask students to collect evidence by recording all the possible combinations of rolls through using *The Game of Pig on a Number Line* Recording Sheet (Reproducible 17). This evidence is one way for students to see why this game is so interesting and engaging.

1. Introduce the recording sheet by first asking students to think about adding the results of rolling two dice when playing *The Game of Pig on a Number Line*:

   "What is the least sum you can roll with two dice?"

   "What is the greatest sum you can roll with two dice?"

   "Can we roll sums for all the numbers from two to twelve?"

2. Display the recording sheet so that everyone can see it. Explain that the top numbers are all of the possible sums you can roll with two dice, starting with the smallest sum, 2, and ending with the largest sum, 12.

3. Now show students how they fill out the sheet. Explain that after each roll, they should record their rolls under the correct sum. Ask one student to roll the two dice and tell the class the numbers they rolled and then the sum of those numbers. Write the expression with two addends under the number on the recording sheet, explaining the process as you do so. For example, if a student rolls a 5 and a 6 say, "Patel rolled a five and a six, the sum of which is eleven, so I'm going to record five plus six under the eleven on the recording sheet."

4. Repeat the previous step for five more sums. This is an example of the recording sheet after five rolls:

2	3	4	5	6	7	8	9	10	11	12
				3 + 3	4 + 3			5 + 5	5 + 6	
					2 + 5					

The Game of Pig on a Number Line Recording Sheet (Reproducible 17) showing five rolls.

Teaching Tip

Pairing Students

Sometimes it is helpful to pair students who have similar abilities so they can share strategies comfortably. If you have students who are second language learners, you would want to pair these students with bilingual partners, if possible. When strong students are paired with struggling students, often the stronger student takes over the game. Find ways to form partnerships with students who are both working within not too broad a range of computational skills.

5. Pair up students and give each pair two dice, a pencil, and a copy of the recording sheet.

6. Ask pairs to take turns rolling two dice until they have each had at least ten rolls each. When the first person rolls, that student tells the other student the two numbers rolled and the sum. The partner writes the two numbers under the corresponding sum on the recording sheet. The students then switch roles, passing the dice and the pencil.

7. After the student pairs have filled out their recording sheets (ten rolls each), bring the class together and lead a whole-class discussion.

8. Use the following questions to prompt discussion around what students notice when they look at the data they have gathered. Although I have provided some sample answers, many of the questions may lead to a variety of answers proposed by you and the students. If, at any time, students seem to be confused, stop the discussion and return to these questions after students have had more experience playing the game.

   - "Why do you suppose we have so few sums of two or twelve?" (*Answer:* "These sums can only be made with one plus one or six plus six. The other sums can be made with other combinations.")
   - "What do you notice about the numbers in the middle of the range?" (*Answer:* "Sums like six, seven, and eight occur more often than sums like two or three or eleven or twelve.")
   - "When do you think you should take another turn in the game of Pig?" (*Answer:* Answers vary. "Maybe you feel lucky?" "Maybe you haven't rolled a one yet and are close to one hundred?" There is no exact answer for this question involving probability.)
   - "When should you stop and pass the dice? Why?" (*Answer:* Answers vary here, too. There is no exact answer for this question involving probability.)
   - "How could we change the game to make it harder or easier to get to one hundred?" (*Answer:* "If we can roll larger numbers beyond twelve, the moves on the number line will be larger. There are dice with eight, ten, and twelve sides that we could use. Or, we could use two spinners with larger numbers instead of two dice.")
   - "What would happen if we used three dice instead of two dice?" (*Answer:* "Getting a one on all three dice at the same time is much less likely than other combinations.")
   - "What if we use a different kind of die or spinners with the numbers zero through nine?" [*Answer:* Answers will vary. "With one spin of two spinners, we might get a sum of eighteen (with nine plus nine) or zero (with zero plus zero).]

Math Matters!

The Frequency of Sums When Rolling Dice

Students who have played games with two dice may have some informal knowledge about the frequency of sums. They often do not get a sum of 2 or 12 because these sums can only be made one way with two dice. Other sums, such as 7, can be rolled with 1 and 6, 2 and 5, and 3 and 4. An addition table for two dice would look like this:

One die could roll each of these numbers: →

The other die could roll these: ↓

+	1	2	3	4	5	6
1	2	3	4	5	6	7
2	3	4	5	6	7	8
3	4	5	6	7	8	9
4	5	6	7	8	9	10
5	6	7	8	9	10	11
6	7	8	9	10	11	12

Addition table for two dice being rolled.

Note that the sum of two ones (1 + 1) occurs only once (see the lightly shaded box). There are ten sums that include a one with another number (see the darker shaded boxes). The information on the table also shows that there are thirty-six different outcomes possible with two dice numbered *1* through *6*. Of thirty-six sums, the probability of rolling a sum of two is one out of thirty-six : $p(2) = \frac{1}{36}$.

Although you do not need to develop this chart or these specific ideas yet with young students, allowing students the opportunity to examine all the ways they actually rolled two dice using Reproducible 17 helps students become familiar with an informal understanding of the uneven likelihood of certain sums with the roll of two dice.

By changing up the game with different kinds of dice, spinners, or setting a larger goal, students can practice working with larger numbers and different combinations of addends while also developing an informal sense of probability.

101 and Out!

Game 6

Overview

In *101 and Out!*, the objective is to be the first player to reach 100 on a number line. A version of this game is also featured in *It Makes Sense! Using the Hundreds Chart to Build Number Sense* by Melissa Conklin and Stephanie Sheffield (Math Solutions, 2012). In that version, students use a hundreds chart as a visual reminder of their progress 0 to 100. In this version, students use an open number line.

To begin, students take turns rolling a die. With each roll, they decide whether the number they roll should be placed in the ones place or the tens place. For example, if a five is rolled, it could be designated as five (ones place) or fifty (tens place). The object of the game is to get to 100 on the number line, *but not beyond 100*. If students' rolls cause them to *pass* 100 on the number line, they are disqualified (hence the game's title: *101 and Out!*). Students must try to reach the objective (getting as close as possible to 100 on the number line) in exactly six turns, by adding the numbers rolled. *Students must use each number rolled.* The game involves the addition of tens and ones and using strategic thinking that deals with place value. It can be played cooperatively or competitively. This is an excellent game for first graders later in the year and for second graders.

Time

40–60 minutes

Materials

open number line labeled *0* and *100,* enlarged for classroom use and demonstration

dice (labeled *1* through *6*), 1 per group of 2 or more students

markers in different colors, 1 per student

101 and Out! Recording Sheet (Reproducible 18), 1 enlarged for classroom use and at least 2 per student

felt placemats for dice rolls (optional)

Related Games

- G-4 Race to 50
- G-6 The Game of Pig on a Number Line

Key Questions

- Will you place the number in the ones place or the tens place? Why?
- What number do you hope to roll next? Why?
- How many more do you need to get to 100 without going over?
- Because you are now halfway through the game, which three numbers do you want to roll to get to exactly 100?
- Which player/pair/team won the game? How do you know? How far away from 100 was the winner?
- What is the difference between your score and your opponent's score?
- What strategy did you use to decide whether to make your roll a group of ones or a group of tens?

Teaching Directions

Introduce

1. Prepare a demonstration open number line labeled *0* and *100*. Display it on the wall at student eye level.

An open number line for use with the game *101 and Out!*

2. Also display the recording sheet (Reproducible 18) so that all students can see it.
3. Gather students on the rug so that everyone has a good view of the number line. Seat them close together to make listening to each other easier.

Time Saver

Preparing the Number Line

The reproducible provided for this game includes space for students to record their numbers on an open number line. Alternatively, distribute whiteboards and black dry-erase markers for students to draw an open number line. You may also laminate copies of an open number line (labeled *0* and *100*) or place copies in plastic sleeves and provide dry-erase markers.

Teaching Tip

Quiet Dice

Rolling dice can create a lot of noise. To lessen the noise, use foam dice or pad students' workspaces with foam or fabric placemats.

Technology Tip

Using an Interactive Whiteboard

If you choose to use an interactive whiteboard, use the interactive pen to draw the open number line on the screen. In addition, you may choose to use the interactive die, so all students can see the number rolled. Once the recording sheet is introduced, you can also scan and display it on the whiteboard.

4. Draw students' attention to the number line and introduce the game. Say, "We are going to play a game called *101 and Out!* It's called *101 and Out!* because the person or team that gets the closest to one hundred after six rolls wins; but, if you pass one hundred and get one hundred one or more, you are out! We'll first play one game as a whole class, and then you'll play with a partner or small group to get as close to one hundred as you can without going over."

5. Model for students how to roll the die. After a number has been rolled, ask the students whether the number should be used as ones or tens. Encourage students to share the reasoning for their decision. For example, let's say the number 5 is rolled. Ask students, "For the first move, do you think you should use the number five or use the number fifty? Do we want to use the ones place or the tens place? Turn and talk to your shoulder partner about this and be prepared to share your thinking with the class."

Example of Student Thinking

Janea: "I don't think we should just go to five because that's not very far. We've got five more rolls to get to one hundred, so we should go to fifty because that's in the middle."

Teacher: "Thumbs up if you agree with Janea's idea." [The teacher then records Janea's thinking.]

Math Matters!

Reading "101"

Reading three-digit numbers and beyond can be confusing for students at the beginning. Often, people say "one hundred and one" for 101. The correct pronunciation is *one hundred one*. In the United States, we use the word *and* to indicate that a decimal point is being used in a number with a fractional part. So, 100.1 is read as one hundred *and* one tenth.

Math Talk

Discussion Buddies

Often, students don't know to whom to turn when teachers ask them to talk to a partner. To save time and alleviate any issues that might arise, assign "discussion buddies" before class discussions. Consider keeping the discussion buddies for several weeks so students become comfortable with their partner, then switch buddies so students have opportunities to work with other classmates.

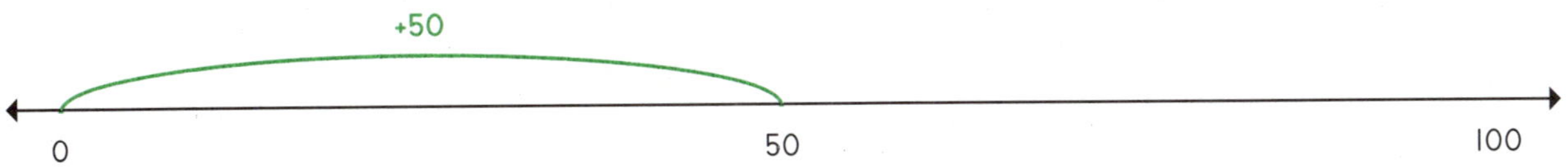

Janea's move after the number 5 is rolled and she decides to assign tens to it rather than ones.

Teaching Tip

Managing the Use of Dice

When rolling dice, sometimes students are overly enthusiastic and throw the dice high in the air or across the room. Set classroom norms for using dice. If using placemats, explain to students that they will lose their turn if they roll the dice off their placemat. Use similar consequences if the dice are not handled responsibly. One way to keep dice safely in play is to put the dice in a small, empty, clear water bottle. This strategy still allows the dice to rattle, but within a contained space. Whatever you choose to do, being proactive in curtailing the misuse of dice makes playing games much safer and more manageable.

6. After you have modeled the appropriate way to roll the die, have a student make the next roll. Once again, ask the class to decide whether the rolled number should be used as ones or tens. In our example, let's say the second roll is a 4. Ask students, "We rolled a four. Should we move four more or forty more? Remember, we only have four more rolls at this point."

Example of Student Thinking

Shannon: "Forty is too big for tens. That will put us at ninety. I think we need to add just four more instead. That will be pretty close to fifty."

Avik: "If we got to ninety, we could get the next ten jumps in four rolls by getting a one and a two and a three and a four." [Some students may argue they could get a six or a five and then be out of the game.]

Teacher: "There are more thumbs up for adding four than forty, so I'm going to record that move, showing small jumps."

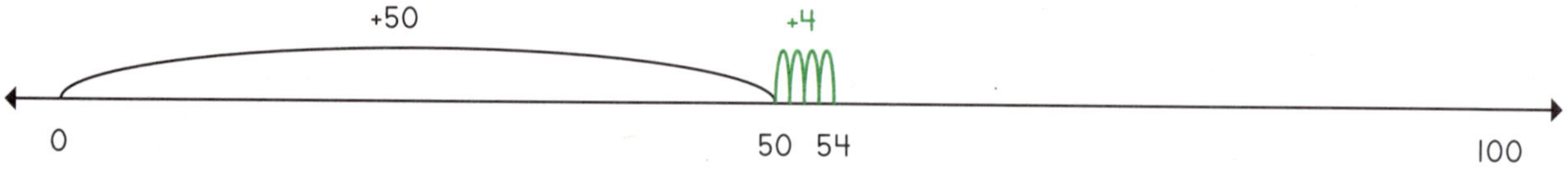

The number line after the number 4 is rolled for the second of six rolls. The students decided to assign ones to it.

Teaching Tip

Modeling

During the game's introduction, the first two rounds of play are ideally used to model the decisions that must be made during the game, whereas the last four rounds engage students in thinking about the decisions. This number of rounds prepares students to play on their own.

7. For the third roll out of six, choose another student to roll the die and ask the class to decide how the rolled number should be used. In our example, let's say the third roll is a 3. Ask students, "Do you think we should make this three or thirty? Explain your thinking to your partner."

Example of Student Thinking

Tanika: "I think we should make it thirty so we'll be at eighty-four. Now we're getting closer to one hundred."

Teacher: "How do you want to make the jump of thirty? By ones or by tens?"

Tanika: "It is easiest for me to make three jumps of ten."

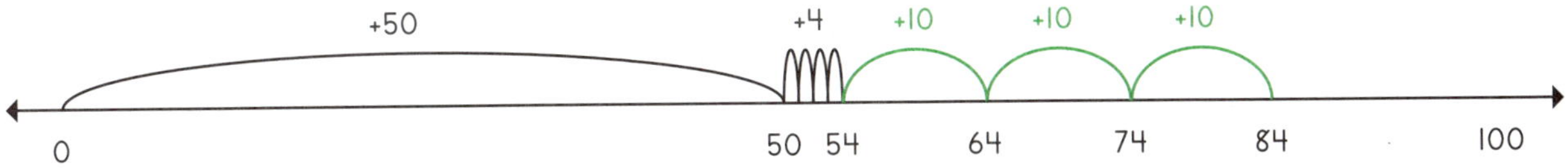

The number line after the number 3 is rolled for the third of six rolls. Tanika decided to assign tens to it.

8. Now that the game is halfway finished, remind students that each player gets six rolls and they've completed three. Pause for a discussion before continuing. Ask, "How many more do we need to get to one hundred without going over?"

Examples of Student Thinking

Jason: "I count from eighty-four to one hundred on my fingers and see that we need sixteen to get to one hundred."

Chinami: "We need six to get to ninety and then ten more."

Jaheem: "Twenty would get us to one hundred four, and that would be too many."

9. Continue with the game, asking another student to roll the die. This time, building on our example, the roll is a 5. Students decide to jump to 89. There is no need for discussion because students are sure that fifty is way too much.

Math Matters!

Probability

Students need many opportunities to roll dice to begin to get an idea of probability. The numbers 1 through 6 are equally likely when one die is used. (Of course, there are some students who think they can roll a 1 if they *think really hard* about how much they need a one for their next round!) Over time, students find that the random nature of this game can make it both exciting and disappointing. These experiences with dice will support student thinking about how probability can change the outcome of a game.

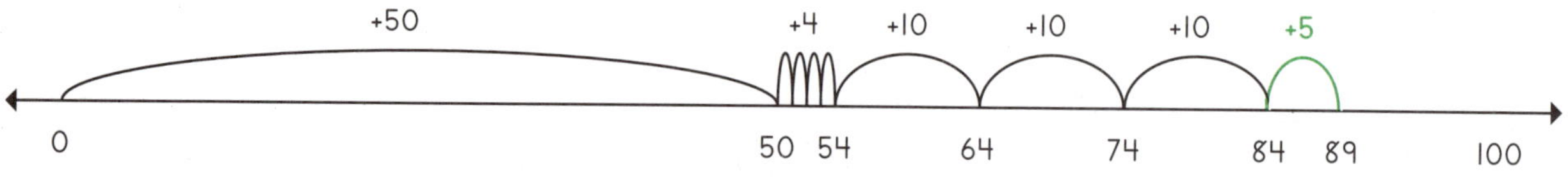

The number line after the number 5 is rolled for the fourth of six rolls. The class decided to assign ones to it.

10. Choose another student to roll the die. Let's say that this time the roll is six and the class agrees to jump to 95. As with the fourth roll, students agree that a jump of sixty would be way too much. Some students may suggest making six into five and one so they can get to 90 and then add five. Most of the students, however, will just count on six more.

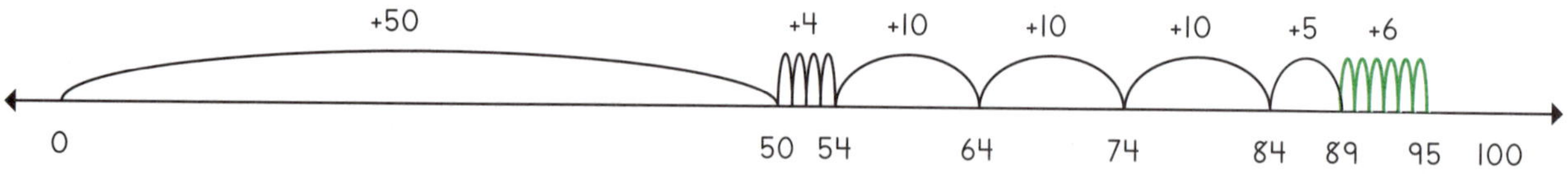

The number line after the number 6 is rolled for the fifth of six rolls. The class decided to assign ones to it.

11. Before making the sixth and final roll, ask students to share their thoughts. Say, "Tell your shoulder partner what the best roll would be for the last number. What would the worst number be?" In our example, most students are sure that a roll of 5 is the best number. Some think a roll of 1 would be the worst whereas others think a roll of 6 is the worst. Remind them that they want to get as close as they can to one hundred without going over, so even a one would be better than a six.

A Child's Mind . . .

Developing Mathematical Habits of Mind

This game encourages the development of the mathematical habit of mind, *reasoning abstractly and quantitatively*. After rolling a number, students need to decide whether to place the number in the ones place or the tens place, taking into consideration the current score and the number of turns remaining.

12. Have a student do the final roll. In our example, the last roll is a 6. Adding six ones lands the class team on 101. Oops! The students are out!

13. Now introduce how to use the recording sheet. Pass out individual copies of the sheet (Reproducible 18). Direct students to the number line at the bottom of the recording sheet. Explain to students that you'll still be working as a class at this point, but they need to record the jumps on their individual number lines after a class decision is made on each rolled number (assigning it ones or tens). Emphasize that the recording sheet is a way for students to keep track of the rolls.

14. Play another round of the game with the whole class. Have them fill out their own recording sheet as the game develops. Simultaneously record the action on a recording sheet posted where everyone can see it. Make sure students also record the moves on the number line. In the example shown the students rolled a 6, 5, and 2 in the first three rolls. For moves, they decided on 6 tens, 5 ones, and 2 tens.

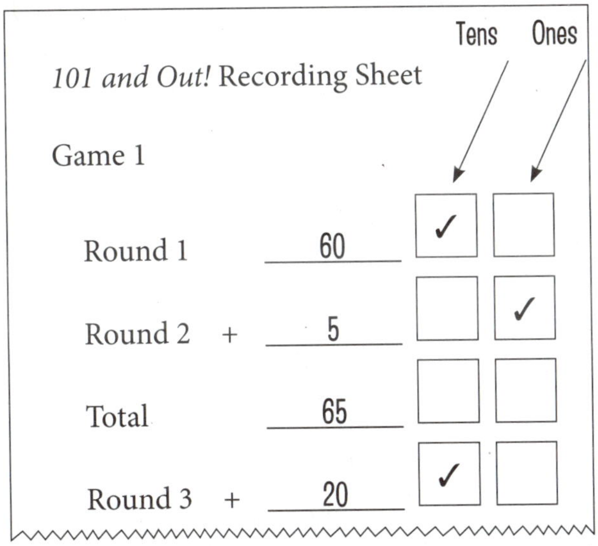

Students record the rolls of 6, 5, and 2 on the recording sheet (Reproducible 18). They decided on 6 tens, 5 ones, and 2 tens.

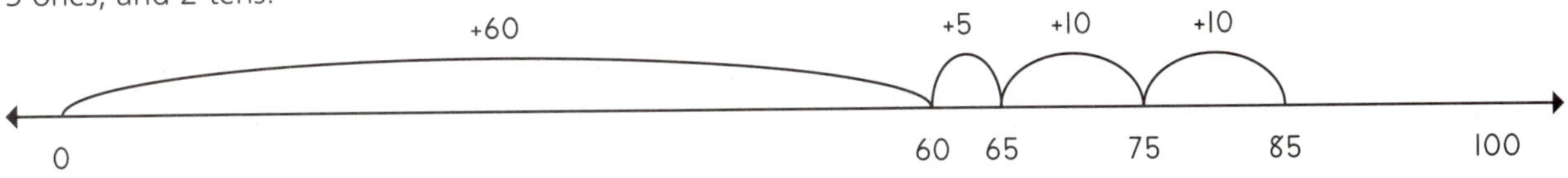

A Child's Mind . . .

Going Over One Hundred

Don't be surprised if students go over one hundred the first several times they play the game. Students need repeated opportunities to learn about magnitude of numbers and number relationships. When students have rolled a number, engage them in a conversation to discuss the two outcomes: What happens if they place the digit in the ones place? The tens place? These conversations help students begin to think strategically about the game.

Explore

15. Now let students play the game independently in groups of two or more. Give each student a new blank copy of the recording sheet (Reproducible 18). Make sure each student has a copy of the recording sheet and a pencil so they can draw their own open number line and keep track of their dice rolls and choices. For more on pairing students, see the Section III introduction on page 166.

16. Remind students of the game's rules: They roll one die, put the number in the ones or the tens place, and record both the move and the sum.

Game Directions
1. Roll the die.
2. Create a number (ones or tens).
3. Record your move on the number line.
4. Find the sum and record it.
5. Pass the die to the next player.

17. Circulate and observe as students play. Ask key questions (see page 216) as appropriate.

Teaching Tip

Playing in Teams

Have students play in groups of four students, two players on each team. That way they can discuss with each other if they want to use their roll as tens or ones. They can also support one another in locating the place on the number line where they write their score so far in the game.

Teaching Tip

Posting the Directions

Posting directions for everyone to see can be helpful for students who are playing the game for the first time. Directions do not have to be detailed. Short, concise phrases help students recall what they should be doing.

Summarize

18. After students have had an opportunity to play independently over a period of time (note that the game can be part of a math workshop learning station), gather them together for a whole-class discussion. The objective will be to link place-value thinking to the recording sheet. Show students an excerpt from a recording sheet as follows. Say, "Here is a recording sheet from an unfinished game that only used three of the six turns."

			Tens	Ones
Round 1		40	✓	
Round 2	+	20	✓	
Total		60	✓	
Round 3	+	5		✓
Total		65	✓	✓
Round 4	+			
Total				
Round 5	+			
Total				
Round 6	+			
Game Total				

19. Ask students to examine the game so far. Say, "Look over the first three rounds and decide how far the score is from one hundred."

Differentiating Your Instruction

Playing in Partners

This game requires students to have a good grasp of magnitude of numbers. Some students realize quickly that putting the number rolled in the tens place creates larger totals; other students need more time to develop this understanding. Students may be more successful at getting close to one hundred without going over if they play with a partner with whom they can discuss their decisions. If you choose to have students play with a partner, then during the introduction to the game, play with a partner so you can model talking to your partner about the choices that arise during the game.

Technology Tip

Using an Interactive Whiteboard

If using an interactive whiteboard, scan the recording sheet and use the interactive pen to fill in the first three rounds. Alternatively, you can display a physical copy of a partially completed recording sheet on the document camera.

20. Now say, "Think about what numbers you want to roll to get to one hundred in the next three rolls." Students could get to 100 with rolls of 5, 1, and 2; 4, 1, and 3; 3, 3, and 2; or any of these groups of numbers in a different order to make up the five needed for the ones place and the thirty needed for the tens place.

21. Ask if any students have a recording sheet from their recent games that shows them getting close to 100 with 6 rolls. Invite them up to share the information on their recording sheets and explain how they were able to use tens or ones to obtain their sums.

22. Ask students to respond to any of the following questions, depending on how much your students seem to understand the strategies of the game.

   - "When do you know that it's a good idea to choose a jump of tens instead of ones?"
   - "What strategy would you use for determining when to make a jump of ones instead of tens?"
   - "How could you get exactly to one hundred with the following rolls: 5, 2, 3, 2, 5, 2? What about 1, 2, 3, 4, 4, 5?"
   - "Is there more than one way to get to one hundred with these same numbers?"

Teaching Tip

Making Connections: Place Value

Some students may find a money representation more meaningful. This game could be played on a money number path with the rolled dice representing the number of jumps of pennies or dimes, with the goal of reaching a dollar, but not more.

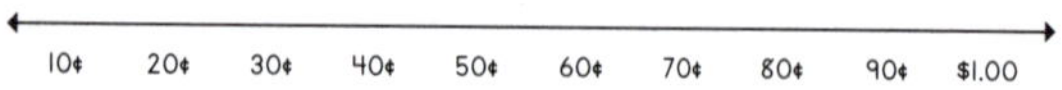

A money number line.

Hot Lava Bridge: Forward and Back, More and Less

Game 7

Overview

This game was invented by imaginative second graders who decided they could move an action figure on a bridge over hot lava (i.e., a number line) to create more fun. Alternating turns, students first draw action cards from a paper bag. The cards determine whether they add or subtract ones or tens on their hot lava bridge/number line. The game ends when one of the players moves the action figure off the number line at either end (hence, falling into hot lava).

This game is an opportunity for students to build fluency and to practice adding and subtracting multiples of tens. As the teacher, you have an opportunity to gather valuable assessment information by asking students, "How do you know where to place the action figure on the number line?"

The game modeled here is most appropriate for first and second graders. The game can also be played in kindergarten. For suggested modifications, see the Differentiating Your Instruction tip in the directions.

Related Lessons

- L-3 Building an Open Number Path
- L-4 Jumping by Ones and Tens

Time

30 minutes

Materials

open number line labeled *0* and *50,* enlarged for classroom use and demonstration

large open number line labeled *0* and *50* (on approximately 18-inch-by-24-inch paper), 1 per group of four students

One Hundred Hungry Ants by Elinor J. Pinczes and Bonnie MacKain (or another favorite counting book)

action figure, chess piece, or other small toy, 1 per group of four students

markers (in different colors), 2 per group of four students

Hot Lava Bridge: Forward and Back, More and Less Action Cards (Reproducible 19), 1 set per group of four students (made from 1 copy of the reproducible)

paper lunch sack, 1 per group of four students (to hold the action cards)

Hot Lava Bridge: Forward and Back, More and Less Recording Sheet (Reproducible 20), 1 per group of four students

Time Savers

Preparing the Number Lines

Instead of making consumable copies of the number line, laminate copies or place them in plastic sleeves and provide dry-erase markers in various colors during game play. You may also have students draw their own number line that is large enough to allow an action figure to move across it.

Preparing the Cards

For the purpose of this game, a set of cards is made with one copy of the reproducible. Copy the action cards on colored paper, cut them out, and place each set into a paper lunch sack.

Key Questions

- How did you know how far to move on the number line?
- Which card would you like to draw next? Why?
- Which card is not a very helpful card to draw? Why?
- How did you know where to write that number?

Teaching Directions

Introduce

1. Select an action figure for demonstrating the game. This might also be an object in the classroom that can be moved easily across a number line.
2. Prepare a demonstration open number line labeled *0* and *50* and place it on the floor. Make sure it is big enough for everyone to see, as well as large enough to move the action figure across it.

An open number line for use with the game Hot Lava Bridge.

3. Gather students on the rug so that everyone has a good view of the number line. Seat them close together to make listening to each other easier.
4. If you have the book *One Hundred Hungry Ants* by Elinor J. Pinczes and Bonnie MacKain (1993), read it with students to present informally the idea of how groups of twos, fives, or tens might be used to create arrangements of items. (If you do not have this book, choose one from your collection that supports this idea.)

Literature Connection

The book *One Hundred Hungry Ants* marches students through a math lesson via rhymed verse and the lighthearted pursuit of a picnic. As the title indicates, one hundred hungry ants proceed single file to a picnic. However, in an effort to get everyone to the food faster, they decide to reorganize—first by dividing into two rows of fifty, then five rows of twenty, and so forth—until their efforts at reorganization cause picnic time to pass them by.

5. Draw students' attention to the number line and explain the game. Say, "We are going to play a game called *Hot Lava Bridge: Forward and Back, More and Less*. The number line represents a bridge over hot lava that our action figure—in this case, a robot—must cross carefully. When the robot jumps off either end of the bridge, the game ends."

6. Model the game. Place a few of the action cards face up so students can become familiar with some of the actions. Say, "Here are some of the action cards. To play, we draw an action card. The number on the action card indicates how far forward (indicated by a plus sign) or how far backward (indicated by a minus sign) we must move our robot."

7. Put all the action cards into a paper lunch sack and draw one. In this example, the card +20 is drawn. Say, "I drew the action card plus-twenty. This means I must move the robot twenty places to the right from zero."

8. Emphasize that students should share their thought process aloud as they move their action figure. In this case, share, "I'm going to put this robot here, where twenty might be on the number line—just a little less than halfway, because I know twenty-five is halfway between zero and fifty."

9. Now emphasize that students need to mark the place their action figure lands by writing the number under the action figure on the number line. Say, "I'm going to write *twenty* under the number line where the robot landed so I know the robot's place on the bridge over hot lava."

Differentiating Your Instruction

Using Fewer Action Cards

For younger students or students who need more practice before working with the full version of this game, limit the action cards being used. I use the following action cards in kindergarten:

+10	−1
−10	+0
+1	−0

I also replace the number line with a number path, with the numbers 1 through 30 written on it (and I don't use the recording sheet).

10. Place the action card back in the paper bag. Explain that, during the game, there are two teams. For purposes of demonstrating the game, you are one team and the whole class the other team. Now it's "Team Students" turn.

11. Have a volunteer student draw an action card for the class. Let's say the student draws the +10 card. Where should the robot move? Ask students to turn and talk with a shoulder partner about the robot's move.

12. Ask students where the robot should be placed. When they indicate the number 30, which is ten more, ask several students to show were they think 30 would appear on the open number line. Have the students signal with thumbs-up if they agree. When there is agreement, choose another student to move the robot to the correct place and have that same student write *30* under the number line.

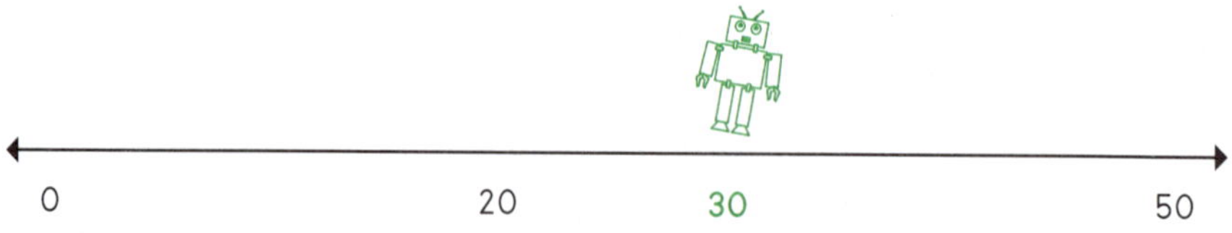

13. Play continues, alternating turns between the class and you until one team moves the action figure off one end of the number line. When this happens, everyone should exclaim, "Hot lava!" and the game ends.

Example Game

+20 (The teacher begins by jumping to 20.)

+10 (The students move the robot to 30.)

-10 (The teacher moves the robot back to 20.)

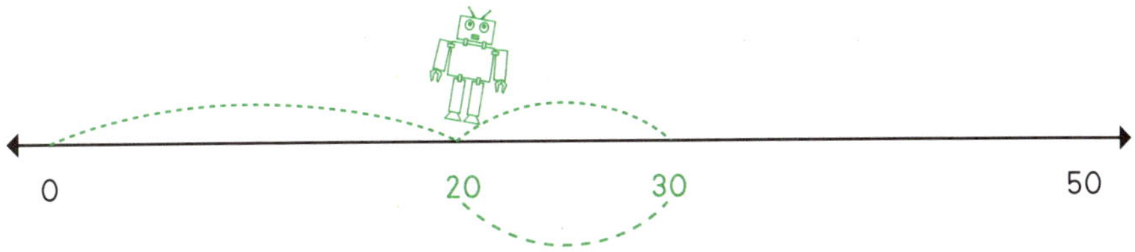

+30 (The students now move the robot to 50.)

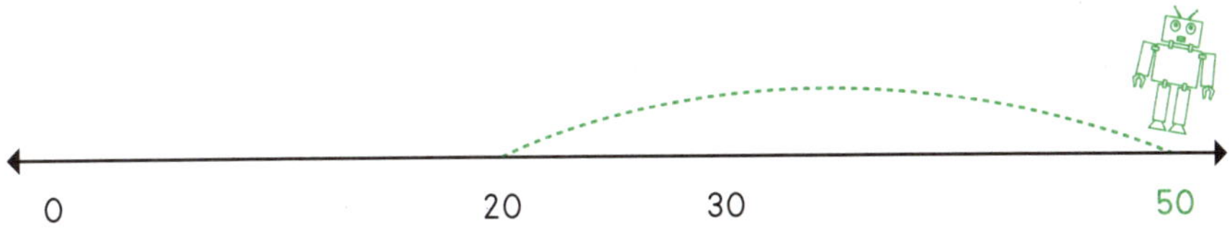

-1 (The teacher moves the robot back to 49.)

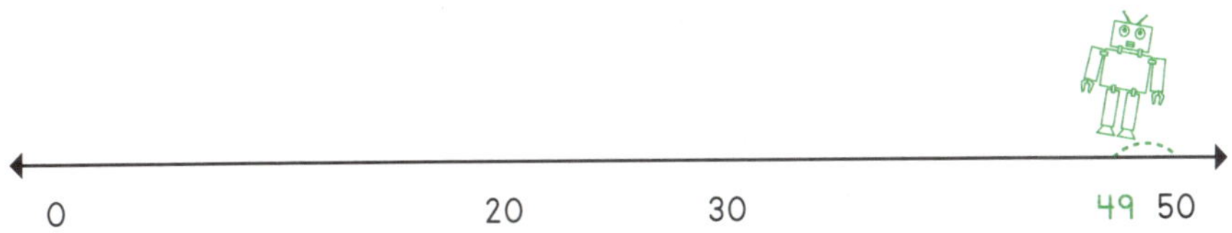

+10 (The students move the robot past 50, to 59, and it falls in the hot lava!)

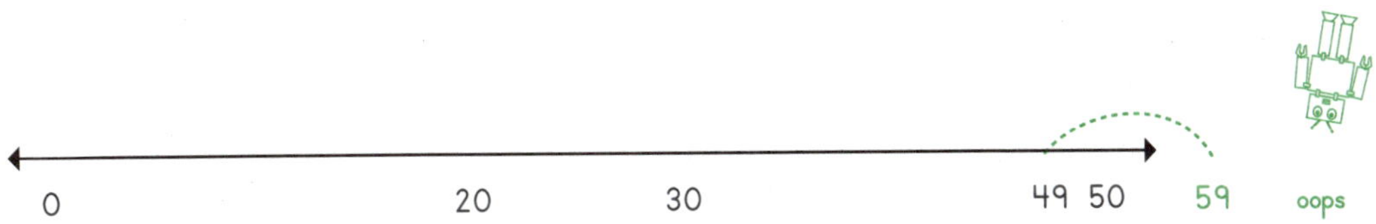

14. After the game has been played with the whole class, draw a number line on the board and show a summary of the moves via jumps. For the example game, the number line would look something like this:

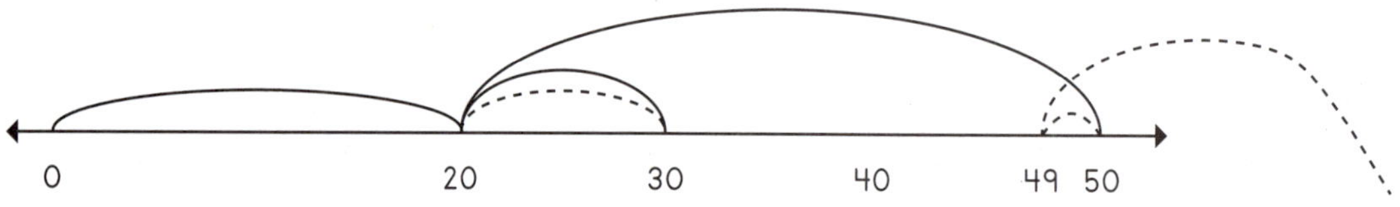

15. Play another round of the game, only this time divide the class into two teams. Introduce the recording sheet (Reproducible 20) and post it where everyone can see it. Explain that students should record the action on the number line as well as the recording sheet. Walk students through a row on the sheet, explaining that each row shows the action of each turn. In the first column, students should write the number where the action figure starts on their team's turn with their own color marker. In the second column, write the number and "the sign" (the operator) on the action card drawn. In the third column, write the number of the new position of the action figure. The second team records their action card and moves on the recording sheet with their own color marker.
16. Facilitate this round of the game as the two class teams play. Clarify directions as needed. Make sure students are recording their moves on both the number line and the recording sheet.
17. After "Hot lava!" happens and the game ends, bring the students' attention to the completed recording sheet. Ask, "What do you notice? What does this recording sheet tell us?"
18. Have students turn to their shoulder partner and talk about their observations, then share as a class.

Math Talk

Discussion Buddies

Often, students don't know who to turn to when teachers ask them to talk to a partner. To save time and alleviate any issues that might arise, assign "discussion buddies" before class discussions. Consider keeping the discussion buddies for several weeks so students become comfortable with their partner. Then, switch buddies so students have opportunities to work with other classmates.

Examples of Student Thinking

Start	Action	End
0	+20	20
20	+10	30
30	-10	20
20	+30	50
50	-1	49
49	+10	59

Deanna: "The robot lost and fell in the hot lava at the end."

Rex: "It always starts at zero."

Saleem: "Plus thirty was the biggest jump."

Keesha: "Minus one was the smallest jump."

19. Ask students to point to where on the recording sheet the robot went backward.

Explore

20. After students are comfortable with the game procedures, give students the opportunity to play the game independently in groups of four, each group divided into two teams of two.

21. Provide each group with a number line (see the Teaching Tip, Preparing the Number Lines on page 225), two different color markers, a lunch bag containing a set of action cards (Reproducible 19), and a recording sheet (Reproducible 20).

22. Allow each group to choose an action figure. If students don't have action figures, have them choose an object in the classroom to use.

23. Remind students of the directions. Say, "When you play this game with your partner, take turns choosing the action card and moving the figure so you both get a chance to do both tasks. Help your partner by

Assessment Opportunities

Fill in the Missing Numbers

For an individual or group assessment, erase or white-out some of the numbers on a recording sheet and ask students to fill in the blanks by figuring out the relationships between one turn and the next. Here is a sample recording sheet (the missing numbers are in the answer key below):

Start	Action	End
0	+20	20
20		30
30	-10	20
20	+30	50
	-1	49
49	-10	
39	-30	9
9	+1	10
	+30	40
	+20	60

Answer Key

Blank in row 2: +10

Blank in row 5: 50

Blank in row 6: 39

Blank in row 9: 10

Blank in row 10: 40

listening to each other's strategy before you move your figure on the number line. Make sure you record each move that matches the action cards. Check each other's work."

24. Help students decide which team goes first. Have each group shake up their paper lunch sack of action cards, then have each team draw a card. The team with the action card that has the greatest number goes first. If they draw a minus card, they can put it back in the bag and draw again.

25. Circulate and observe as students play. Ask key questions (see page 226) as appropriate.

Summarize

26. After students have had an opportunity to play the game independently over a period of time (note that the game can be part of a math workshop learning station), gather them together for a whole-class discussion. Show the class a recording sheet from a previous game. Explain, "I saved one of the recording sheets from our games. Look at the recording sheet and discuss with your shoulder partner what happened in the game."

27. After partner discussions, bring students back together. Cover up the recording sheet, revealing only the first row. Ask students, "What happened on the first turn of this game?" Have students state their thinking (or act it out with the action figure).

28. Continue to have students explain their thinking, revealing each row accordingly.

Extend Their Learning!

Drawing Two Cards

To increase the challenge of this game, have students draw two cards at a time and combine the numbers into one move on the number line. For example, if a student draws the action cards +20 and –10, the move would be +10 on the number line.

Example of Student Thinking

Shinmin: "I got a minus one and a plus twenty, so I'm going to move nineteen in this direction. I started at ten, so the robot's going to be at twenty-nine."

Tomas: "I got a plus ten and a minus ten, so the robot stays in the same place, at twenty-nine."

Deanna: "This is a minus ten and a plus thirty, so that's twenty more. Now the robot's at forty-nine. Look out!"

Game 8 The Larger Difference Game

Time

30 minutes

Materials

2 colored pencils or markers (1 red and 1 blue), 1 set for each group of 4 students

dice (labeled *1* through *6*), 2 for each group of 4 students

The Larger Difference Game Recording Sheet (Reproducible 21), 1 copy for each group of 4 students

Greater Than, Less Than, and Equal To Reference Chart (Reproducible 2), 1 copy enlarged and displayed on wall

Overview

In *The Larger Difference Game*, opposing teams roll two dice labeled *1* through *6* and create two, two-digit numbers from the roll. They are then tasked with finding the difference between the two numbers, using an open number line to support their thinking. Teams compare their differences: Which difference is larger? Teams jointly fill out a recording sheet for each round of the game. The team with the larger difference is considered the winner of that round.

This collaborative process supports student use of the symbolic representation of the greater than sign. Because this game is played on an open number line, students have a chance to show jumps in ways that match their level of understanding. This game is most appropriate for first and second graders.

Related Lessons

- L-4 Jumping by Ones and Tens
- L-5 Jumping by Ones, Fives, Tens, and Twenty
- L-9 Solving Comparison Problems

Key Questions

- What two numbers can you create with these digits?
- How would you determine the difference between these two numbers on the number line?
- Who has the larger difference? How do you know?
- How did you keep track of the jumps to find the difference?
- What do you notice about all the differences you are finding?

Teaching Directions

Introduce

1. Draw two open number lines on the board. Label one *Red Team* and one *Blue Team*. Gather students together so they can see the number lines easily.

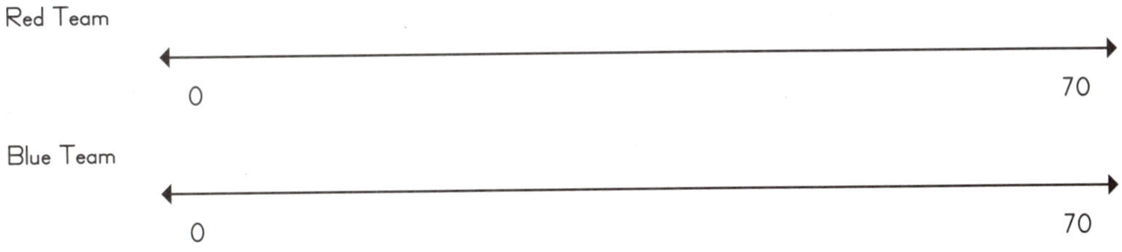

Open number lines for use with this game *The Larger Difference Game*.

2. Let students know they will be playing a game called *The Larger Difference Game*. It will help them strengthen their skills in finding the difference between two numbers. Explain there are two teams: a red team and a blue team. Divide the class so one side is the red team and one side is the blue team. Give each team a corresponding colored pencil or marker.

3. Ask for a volunteer from the red team to roll two dice and report the numbers. Explain that these numbers are the digits for their first two numbers in Round 1 of the game. For example, if 2 and 4 are rolled, the numbers are 24 and 42.

4. Emphasize one rule: If students roll the same number on each die, they need to roll the dice again so that both numbers are different.

5. Display an enlarged version of the recording sheet (Reproducible 21). Show students how to record their numbers on the recording sheet. Repeat with the blue team. (In the following example, the red team rolled a 2 and a 4; the blue team rolled a 3 and a 6.)

A Child's Mind . . .

Math Vocabulary

Math vocabulary like "digits" is best introduced within a successful learning context such as *The Larger Difference Game*. When children hear vocabulary over and over again within an appropriate context, they are able to make sense of it and begin to understand the meaning. Just like young students learn the word *Tyrannosaurus Rex*, they can learn math words like *digits* and *numerals*.

Round 1

The red team numbers are _24_ and _42_.
The difference between these numbers is
________.

The blue team numbers are _36_ and _63_.
The difference between these numbers is
________.

The team with the larger difference: Red Team   Blue Team (circle one)
_____ > _____.

6. Explain that students will find the difference between their two numbers using an open number line to help them. On the board, using two different-colored markers, help students decide where to place the numbers on their team's open number line. Say, "Our line is just to help us remember the numbers we are using. Our lines don't have to be perfectly straight or measured."

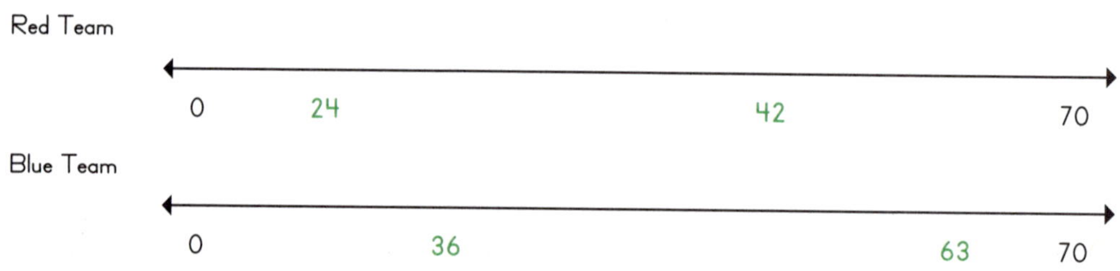

The teams' number lines showing the numbers from their first rolls.

Teaching Tip

Think-Pair-Share

When, as teachers, we ask a question to the whole class, often one student responds. However, when we use the think-pair-share strategy, all students get a chance to talk about the question. This can be a chance for students to rehearse with a partner what they may want to say to the whole class, or an opportunity for them to predict what will happen in a safe setting.

7. Ask students "How would you determine the difference between the numbers on your team's number line?" Remind students they can jump from the smaller number to the larger number or jump from the larger number to the smaller number.

8. Do a think-pair-share, then record a student's thinking from each team. Check with students to ensure your recording represents their thinking correctly.

Examples of Student Thinking

Example A1

Camden (Red Team): "You could jump one to twenty-five, then five to thirty, then ten to forty, and then two more to forty-two. So that's eighteen jumps. The difference is eighteen." [The teacher records what Camden says.]

Teaching Tip

Recording

As you record students thinking, make sure to check with them to confirm your drawing represents their thinking correctly.

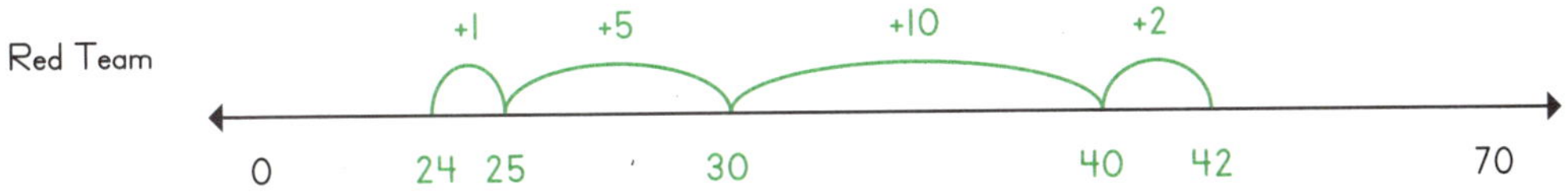

Number line showing Camden's thinking in determining the difference of 18 between the red team's numbers 24 and 42.

Example A2

After the teacher records what Camden says, the teacher asks, "Where do we see the difference of eighteen?"

Camden: "You add the one and the five and the ten and the two to get eighteen."

A Child's Mind . . .

Finding Shortcuts

At first, students may need to draw jumps on the number line one step at a time. After several turns or games, students may find shortcuts and be able to jump multiples of tens. Encourage this strategy and ask them to explain how they know they landed on and recorded the correct number.

Example B

Kiko: "I got eighteen, too. You could jump six to thirty and then twelve more to forty-two and six and twelve is eighteen." [The teacher records what Kiko says.]

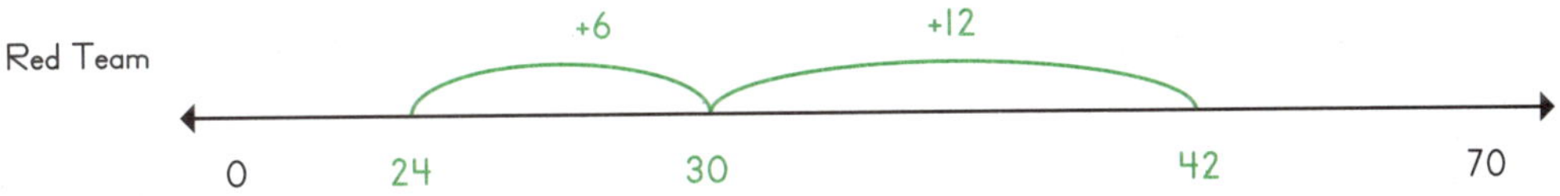

Number line showing Kiko's thinking in determining the difference between the red team's numbers 24 and 42.

Example C

Pat: "You could go to thirty-four. That's ten more. Then go to forty-four; that's another ten, so that's twenty. Then you have to go back two to forty-two, so it's going to be eighteen because you are going back."

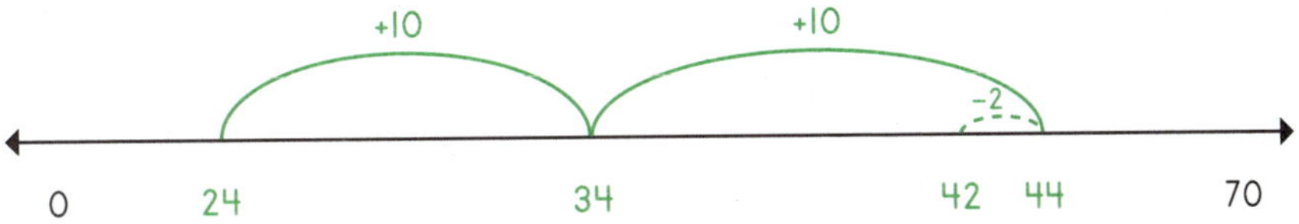

Number line showing Pat's thinking in determining the difference between the red team's numbers 24 and 42.

Math Matters!

What Are the Differences?

With two dice labeled *1* through *6*, there are only fifteen possible rolls that result in two different digits.

Remember, in this game, if the roll results in the same two digits, the team must roll again. The rolls are as follows:

1, 2	2, 3	3, 4	4, 5	5, 6
1, 3	2, 4	3, 5	4, 6	
1, 4	2, 5	3, 6		
1, 5	2, 6			
1, 6				

In each case, when the two numbers are formed from the two digits rolled, the differences between the two numbers are nine, eighteen, twenty-seven, thirty-six, or forty-five—all multiples of nine.

Math Matters!

The Equal Sign

When red and blue teams compare their differences, they may discover their differences are equal. In this case, they need to write an expression using the equal sign. Some children may have never seen the equal sign used in this way (for example, 18 = 18 or 36 = 36). They may be used to seeing it used only in equations, with two addends on one side and the sum on the right, as in 9 + 18 = 27. Some children think of the equal sign as a kind of "punctuation" that shows the end of a number sentence instead of thinking about it as a sign that expresses a relationship. Take time to discuss this with students.

Some children may also think that an equation such as 18 = 42 – 24 is not true because there is only one number on the left and two numbers on the right. With more exposure and explicit conversation about the meaning of the equal sign, students eventually become more comfortable with equations such as 4 + 4 = 5 + 3, in which there are two or more addends on each side of the equal sign.

Teacher: "Could you say more about how you changed twenty to eighteen in the second part?"

Pat: "Well, I like to jump by tens and forty-four is close to forty-two, but too far. So I had to take two jumps away because two jumps of ten were too many." [The teacher records what Pat says.]

9. Call on a volunteer from the blue team to share with the class, using some of the previous modeling strategies.
10. Return to the recording sheet and show the class how to enter the differences in the appropriate places:

> The red team numbers are _24_ and _42_.
> The difference between these numbers is _18_.
>
> The blue team numbers are _36_ and _63_.
> The difference between these numbers is _27_.
>
> The team with the larger difference: Red Team Blue Team (circle one)
> _____ > _____.

11. Now ask, "Which team has the larger difference?" Instruct the students to talk to their partner.
12. Ask for a volunteer from the team with the largest difference (in our example, the blue team) to come up to the recording sheet and circle the team with the larger difference.
13. Write the corresponding expression on the recording sheet. In our example, the blue team's difference, 27, is greater than the red team's difference, 18, so the expression is 27 > 18.
14. Read the expression aloud. As needed, reference the Greater Than, Less Than, Equal To Reference Chart (Reproducible 2).

15. If the two differences are equal, write an equation using an equal sign instead of the greater than symbol (see the Math Matters! tip for a description of this situation).

16. Facilitate several more rounds with new volunteers from the red and blue teams until students show confidence about how to play the game.

Explore

17. Divide students into groups of four to play the game independently. Provide each group with a red marker and a blue marker, two dice, and a recording sheet (Reproducible 21).

18. Have the groups of four divide into two teams of two (a red team and a blue team) and roll the dice to determine which team goes first (the team with the larger roll starts). Suggest that students take turns recording with their partner. When one student is recording, the other rolls the dice.

19. Circulate around the room and support students when necessary by asking key questions (see page 234).

Summarize

20. After the small groups have had the opportunity to play the game independently over a period of time (note that the game can be part of a math workshop learning station), gather them together for a whole-class discussion. Choose some of the recording sheets to share as a class. Display the recording sheets where everyone can see them.

21. Ask students to look at the recording sheets and see whether other groups used any of the same numbers they used. Ask, "Did you find the difference between the two numbers in the same way as your classmates?"

Teaching Tip

Recording Sheets

Save the students' recording sheets because the sheets give insights on how students are thinking about making moves on a number line.

22. Have two or three students volunteer to explain how they jumped from one number to another in a different way than one of their classmates.
23. Summarize by reminding students that the purpose of the game is to find the differences between two numbers (in any way that makes sense) and determine which difference is larger.
24. Save the recording sheets and post the following number sentences from them on the board:

    61 – 16 = 45

    62 – 26 = 36

    63 – 36 = 27

    64 – 46 = 18

    65 – 56 = 9

25. Ask students to look at these results from their games. Explain, "These were the scores when one die was a 6. "What patterns do you see? What do you notice?"

Examples of Student Thinking

Carina: "The ones numbers go in order going down. Sixty-one, sixty-two, like that."

Jenny: "The answers go down in the tens and up in the ones."

Marty: "There's six in the first number and another six in the second number."

Teacher: "You are all noticing some interesting patterns."

Extend Their Learning!

Drawing Two Number Cards 1 through 9

In this extension, teams use number cards labeled *1* through *9* (see Reproducible 1) instead of the two dice. (Reproducible 1 includes cards numbered *1* through *20*. Use cards 1 through 9 only from this reproducible.) Place the cards in a paper lunch sack. The first team draws two numbers from the sack and forms two, two-digit numbers. They write the two-digit numbers on their recording sheet. Then they place the numbers back in the sack and the second team draws, forms, and records the numbers. This version gives students the opportunity to work with a larger range of numbers, such as ninety-two and twenty-nine, but still have differences that are multiples of nine.

Using Three Dice Instead of Two

In this version, use three dice instead of two dice to compare two, three-digit numbers. It is possible, with one roll of three dice, to create six different three-digit numbers. So, in this version, students use number sense and estimation to decide how to determine the place values for the three digits to obtain the largest possible difference. For example, if the three digits are two, six, and five, the difference between two hundred fifty-six and six hundred fifty-two is the greatest difference possible with these numbers.

Consider adding the rule that if two of the three dice rolled have the same number, players must roll the three dice again until all three are different numbers.

Examples of Student Thinking

Teacher: "Mona has rolled a two, a five, and a six. What two numbers can you create with these three digits?"

Hakeem: "We could have two hundred sixty-five and five hundred sixty-two."

Teacher: "Are there other numbers you could make that have a greater difference?"

Loki: "If we had a number in the two hundreds and the other number in the six hundreds, we'd have a difference of four hundred."

Teacher: "It sounds like you have an idea about how to make two numbers, where one is the smallest and the other is the biggest."

Loki: "Can we use the blocks?" [referring to the place-value blocks]

Teacher: "Use the tools you need to make your number. [At this point, the members of the red team use the blocks to create the greatest number because it has the most wood! Then they create the least number using these same digits and report that this number has less wood.)

Teacher: "Let's let the blue team roll their dice and make their number."

Teacher: "We can see that Avik has rolled a four, a one, and a six. What two numbers can you create with these digits?"

Brendan: "We could have four hundred sixteen and one hundred sixty-four."

Teacher: "Are there other numbers you could make that have a greater difference?"

Jaleel: "One of our numbers is going to be six hundred forty-one. The other is going to be one hundred forty-six because that one is the smallest."

[The teacher then uses the recording sheet to show the two team numbers. Mona, on the red team, explains their thinking.]

The red team numbers are 256 and 652.
The difference between these numbers is ___.

The blue team numbers are 641 and 146.
The difference between these numbers is ___.

Mona: "We jumped to two hundred sixty, and that was four. Then we jumped forty more to three hundred, and that was forty-four so far. Then we jumped three hundred to six hundred. That was three hundred forty-four. Then we jumped fifty more to six hundred fifty, so that's three hundred ninety-four so far. Then we jumped two more to six hundred fifty-two. So we jumped three hundred ninety-six jumps all the way, which is close to our idea of four hundred."

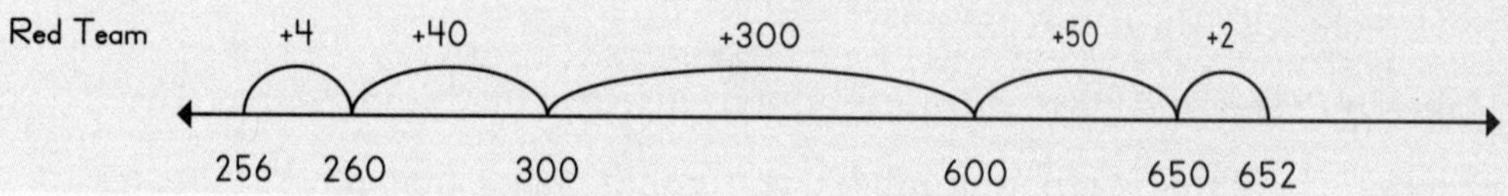

$$256 + (4 + 40 + 300 + 50 + 2) = 652$$

$$256 + 396 = 652$$

[Avik, from the blue team, then shares the blue team's recording and explains their thinking.]

Avik: "We jumped four to one hundred fifty, then fifty to two hundred, then four hundred to six hundred, then forty and one more."

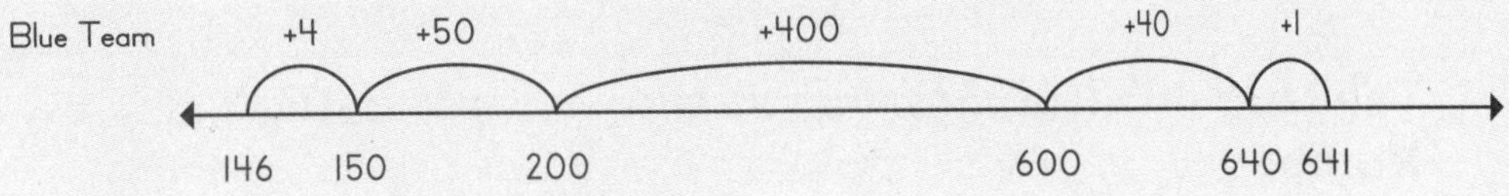

$$146 + (4 + 50 + 400 + 40 + 1) = 641$$

$$146 + 495 = 641$$

Teacher: "Now look at your recording sheets and determine which team has the largest difference. Explain how you know this."

Students were initially confused about the idea of the greatest number 652 versus 641 compared to the largest difference between two numbers that they could make with the three digits. It may be that the three-digit numbers with which they are working are not as friendly to them as the two-digit numbers they know are between one and one hundred.

After students have shared their explanations, return to the recording sheet to determine the winning team:

The red team numbers are <u>256</u> and <u>652</u>.
The difference between these numbers is <u>396</u>.

The blue team numbers are <u>641</u> and <u>146</u>.
The difference between these numbers is <u>495</u>.

The team with the larger difference: Red Team (Blue Team) (circle one)
<u>495</u> > <u>396</u>

To summarize this version of the game, ask students to describe other ways they could determine the differences between their two numbers using different-size jumps. If other combinations of the three digits could have been used to create a greater difference, prompt students to try those combinations.

Drawing Three Number Cards, 1 through 9

Instead of rolling three dice, place number cards labeled *1* through *9* (see Reproducible 1) in a paper lunch sack and have teams each draw three cards, then form three-digit numbers. Proceed similarly to the previous extension.

Teacher Reflection

The Larger Difference: Subtraction with Two-Digit Numbers

With this game, I got a chance to see how second graders determined the difference between two, two-digit numbers. Per my suggestion, students used an open number line instead of their usual method of subtraction. They had been taught how to subtract using a regrouping algorithm: Look at the numbers in the ones place, decide whether to go to the tens place and cross out that number, write down one less in its place, and put a little one next to the number in the ones place. Sometimes, however, students would forget to do this and subtract the smaller digit from the larger digit, regardless of whether the problem required regrouping. When they were presented with a lot of problems that did not require regrouping, this method worked fine; but with the two-digit numbers, it was hard for some of them to know when to regroup.

I wondered whether they knew that the numbers were equivalent when they regrouped them (meaning, for example, 30 + 12 is the same as 42). I wanted to help students choose to use procedures that made more sense than the ones they were using. I hoped that the number line game *The Larger Difference Game* might provide the context they needed.

After we had played the game for several days, I gathered students around the whole-group area and showed them a sample recording sheet from the game:

Round 1

The red team numbers are 25 and 52.
The difference between these numbers is 27.

The blue team numbers are 37 and 73.
The difference between these numbers is 36.

The team with the larger difference: Red Team (Blue Team)
(circle one) 36 > 27

I asked a member of the blue team to show us how he obtained the difference of thirty-six on the number line. He did a nice job of explaining the difference by jumping from thirty-seven to forty and then jumping thirty more to seventy. Then he jumped the last jump of three to seventy-three.

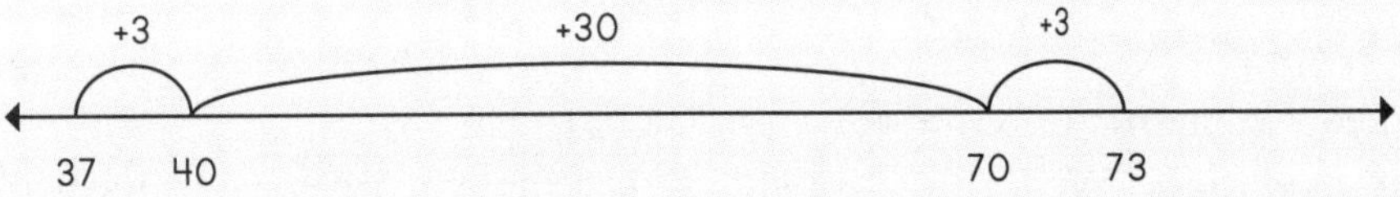

I asked the red team to show us how they got their difference of twenty-seven and they also were able to do this on a number line in much the same way.

Then I showed students these two subtraction problems from a worksheet they had used earlier in math class:

$$\begin{array}{r} 52 \\ -25 \\ \hline 33 \end{array} \qquad \begin{array}{r} 73 \\ -37 \\ \hline 44 \end{array}$$

I explained to the class that the student who did these problems told me that she took two away from five and two away from five in the first problem, and three away from seven and three away from seven in the second problem. I asked the class to think about what the student said.

One student noticed that the numbers in these two problems were the same as the numbers in *The Larger Difference Game* Recording Sheet. We confirmed this was so by comparing the two sets of problems. Then we looked at the two different answers and Olivia said, "Sometimes you get a different answer if you do the problem a different way."

I asked the class to think about what Olivia said. "Is this possible? Can you get two different *right* answers?" They seemed confused about this question.

Stephanie noticed there was an error in the method used in the second group of problems. "I think she shouldn't take two away from five. She needed to change the two into twelve and regroup," Stephanie stated.

I responded, "Do you think you could do these problems with a regrouping strategy?"

We moved to the board and did the subtraction again, using the regrouping strategy shown here:

$$\begin{array}{r} {\scriptstyle 4\ 12} \\ 52 \\ -25 \\ \hline 27 \end{array} \qquad \begin{array}{r} {\scriptstyle 6\ 13} \\ 73 \\ -37 \\ \hline 36 \end{array}$$

Mona said, "See? The answer is the same now, so I think they're both right."

I replied, "Yes, I think you are correct. I wonder why that student made those errors?"

Olivia said, "I think it's because subtraction is hard and you want to just get the answer."

I asked, "What makes it hard?"

She responded, "Well, the numbers are too big, so you have to look at the tens and ones."

I asked the class if someone else had an idea about why the errors were made. Another student said, "I think sometimes you're not careful and just make a mistake." Others agreed with this.

I asked, "Is there some way we could remember to be more careful?"

Brendan said, "I think you should check your work, like add them to see if they're right." (He was referring to the strategy of adding the difference to the subtrahend to see if the total matches the minuend.)

I asked Brendan, "Can you explain what you mean or why this works?"

Brendan answered, "Well, it's like the number family, when you take away ten minus four and get six, then you add six and four and you get ten."

I thanked Brendan for the explanation and we tried his method to see if thirty-seven plus thirty-six was seventy-three. It worked!

I hope that with more experience with the open number line these students use it confidently instead of their error-prone regrouping strategy. It seems like they have a much better chance of making sense of the size of the numbers with which they are working when they use the open number line, instead of just looking at the digits in the ones or the tens place. (When they jumped from thirty-seven, for example, they were still thinking about more than just the seven in thirty-seven.)

My challenge is to see whether students are more comfortable with this method than the one they were "taught" in their workbook. Seeing how students responded to this discussion was an eye-opening opportunity for their main classroom teacher. She saw how they were beginning to make sense of the numbers using the open number line.

The Smaller Difference Game

Game 9

Overview

In this game, opposing teams roll two dice labeled *1* through *6* and create two, two-digit numbers from the roll. They are then tasked with finding the difference between the two numbers using an open number line to support their thinking. Teams compare their differences: Which difference is smaller? Teams fill out a recording sheet together for each round of the game. The team with the smaller difference is considered the winner of that round.

This collaborative process supports student use of the symbolic representation of the *less-than* sign. Because this game is played on an open number line, students have a chance to show jumps in ways that match their level of understanding. Furthermore, students may find it interesting that the winner for each round of the game is the team that has the *smaller score,* not the larger score.

It is recommended that students play this game on a different day than *The Larger Difference Game* (G-8) to help them focus specifically on smaller differences. This game is most appropriate for late first graders and second graders because they will be determining differences between a pair of two-digit numbers, typically a second-grade challenge.

Time

30 minutes

Materials

large open number line labeled *0* and *70* (on approximately 12-inch-by-24-inch paper), 2 per group of four students

colored pencils or markers (red and blue), 1 set for each group of 4 students

dice (labeled *1* through *6*), 2 for each group of 4 students

The Smaller Difference Game Recording Sheet (Reproducible 22), 1 copy for each group of 4 students

Greater Than, Less Than, and Equal To Reference Chart (Reproducible 2), 1 copy enlarged and displayed on the wall

Related Lessons and Game

- L-4 Jumping by Ones and Tens
- L-5 Jumping by One, Fives, Tens, and Twenty
- L-9 Solving Comparison Problems
- G-8 The Larger Difference Game

Key Questions

- What two numbers can you create with these digits?
- How would you determine the difference between these two numbers on the number line?
- Who has the smaller difference? How do you know?
- How did you keep track of the jumps to find the difference?
- What do you notice about all the differences you are finding?

Teaching Directions

Introduce

1. Draw two open number lines on the board. Label one *Red Team* and one *Blue Team*. Gather students together so they can see the number lines easily.

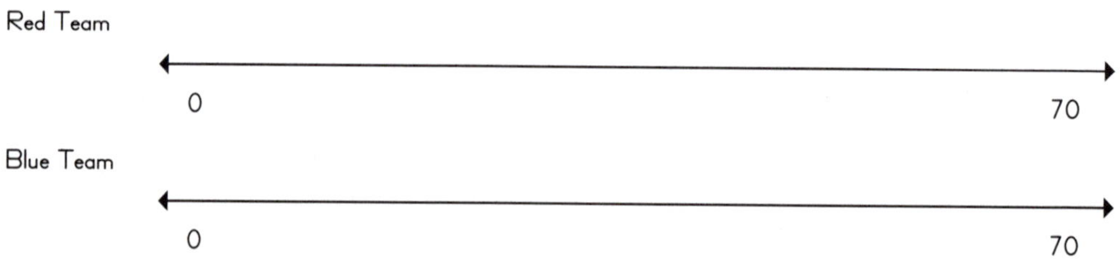

Open number lines for use with *The Smaller Difference Game.*

2. Let students know that they will be playing a game called *The Smaller Difference Game.* It will help them strengthen their skills in finding the difference between two numbers. Explain there are two teams: the red team and the blue team. Divide the class so one side is the red team and one side is the blue team. Give each team a corresponding colored pencil or marker.

3. Before introducing the use of the dice, model a sample round with students using the following predetermined numbers:

    Red team: 10, 25

    Blue team: 10, 30

    Explain that each team needs to place its team's numbers on the corresponding number line on the board. Ask students first to think about where the number 25 would appear on the top number line. This helps them figure out where to place the 30 on the bottom number line.

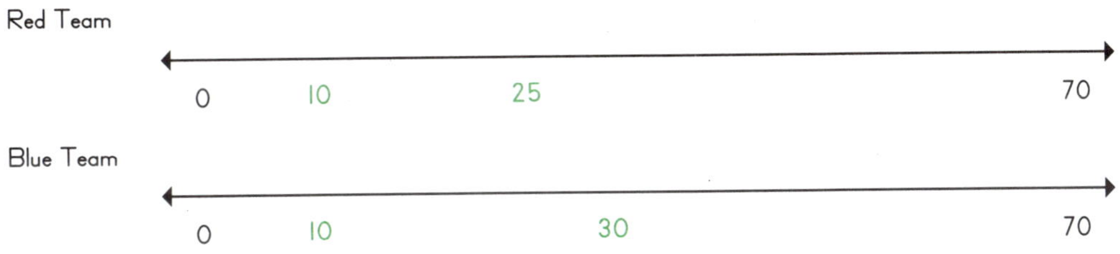

Each team's number line labeled with the predetermined numbers.

4. Now ask the class, "Which number line do you think has the smallest difference? Why?"

    First, draw students' attention to the red team's number line. Ask, "How can someone on the red team tell us how far it is from ten to twenty-five?" Record student thinking on the number line.

Example of Student Thinking

Jordan: "Jump fifteen more and you get to twenty-five because fifteen is five and ten."

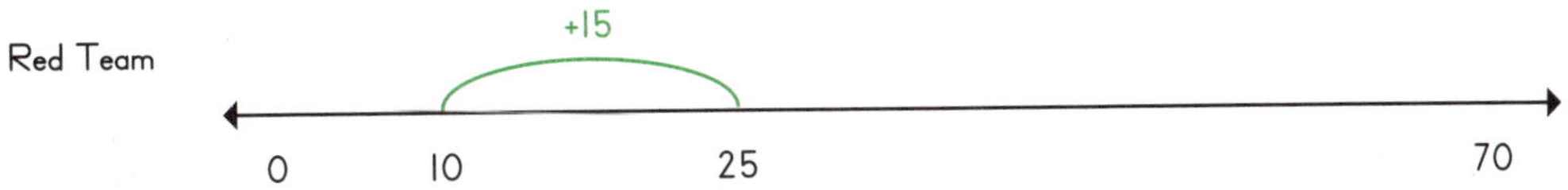

The red team's number line showing the thinking for finding out the difference between the numbers.

Now focus on the blue team's number line. Say, "I can see that the difference between ten and thirty is greater. Who on the blue team can tell how to find the difference for ten and thirty? Record student thinking on the blue team number line.

Example of Student Thinking

Suejin: "Thirty is twenty more than ten because when you count by tens you say *ten, twenty, thirty*."

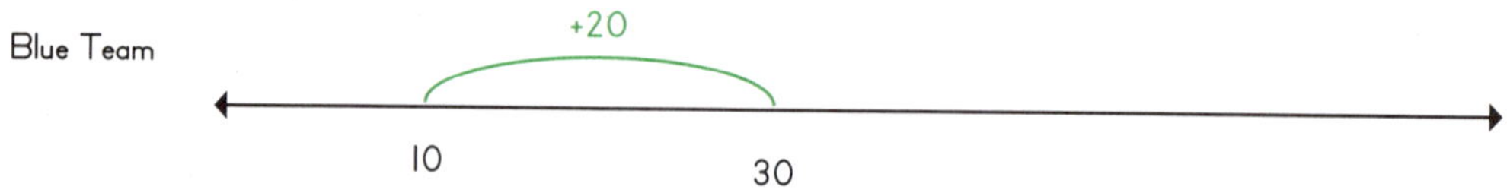

The blue team's number line showing the thinking for finding out the difference between the numbers.

5. Ask again, "Which difference is smaller or less?" In this example, students should respond, "Fifteen is smaller than twenty."
6. Record students' thinking as an expression using the less-than sign:

   $15 < 20$

7. Say, "Fifteen is less than twenty, so fifteen is the smaller difference. Do our two number lines show this thinking accurately?" Now introduce the objective of the game: The winning team is the team with the smaller difference. Emphasize the word *smaller* when saying this with gestures for *smaller*. (I use a gesture of holding a ball with two hands that is being blown up to show *more*. When I want to gesture *smaller or less*, I bring my hands closer together as if holding a balloon or ball that is losing air.)

Teaching Tip

Less and More

Gestures are very helpful in supporting students' understanding of *less* and *more*. Using the open number line, you can also show *less* by spreading your hands over the difference noted in the top number line, then spread your hands over the difference in the bottom number line to indicate *more*.

8. If students have played Game 8, *The Larger Difference Game*, ask students to explain in their own words the difference between this game and that one.

Examples of Student Thinking

Sylvana: "You want to find a difference between two numbers."

Teacher: "Yes, but what does *a smaller difference* mean?"

Rose: "Instead of finding a bigger difference, you want to find the smallest one."

Teacher: "Yes, in this game we want to find the smaller difference. If you had differences of nine and eighteen, which is the smaller difference?"

Ahmed: "Nine is smaller than eighteen."

9. Support students' explanations with comments such as, "Smaller means less. So, if a difference of nine is compared to a difference of eighteen, we can say nine is less than eighteen."
10. Now share an enlarged version of the recording sheet (Reproducible 22). Begin to play the game. Ask for a volunteer from the red team to roll two dice and report the numbers. Explain that these numbers will be the digits for their first two numbers in Round 1 of the game. For example, if the red team rolls a 1 and 5, their numbers are 15 and 51.
11. Emphasize one rule: If students roll the same number on each die, they need to roll again.
12. Show students how to record their numbers on the recording sheet.

A Child's Mind . . .

Math Vocabulary

Math vocabulary words such as *digits* are best introduced within a learning context—in this case, in *The Smaller Difference Game*. When children hear vocabulary over and over again within an appropriate context, they are able to make sense of it and begin to understand the meaning. Just like young students learn the word *Tyrannosaurus rex*, they can learn math words such as *digits* and *numerals*.

13. Ask for a volunteer from the blue team to roll two dice and report the numbers. In our example, the blue team rolls a 3 and a 2. Record their numbers. The recording sheet now looks like this:

> **Round 1**
>
> The red team numbers are _15_ and _51_.
>
> The difference between these numbers is ________.
>
> The blue team numbers are _23_ and _32_.
>
> The difference between these numbers is ________.

14. Now direct students' attention to the number lines on the recording sheet. Explain that they will use number lines to help them determine which team has the smaller difference.

15. Model the use of the number line using the red team's numbers first. Say, "Fifteen is closer to the beginning of the number line whereas fifty-one is farther away. We want to know the difference between these two numbers." Mark the numbers on the number line.

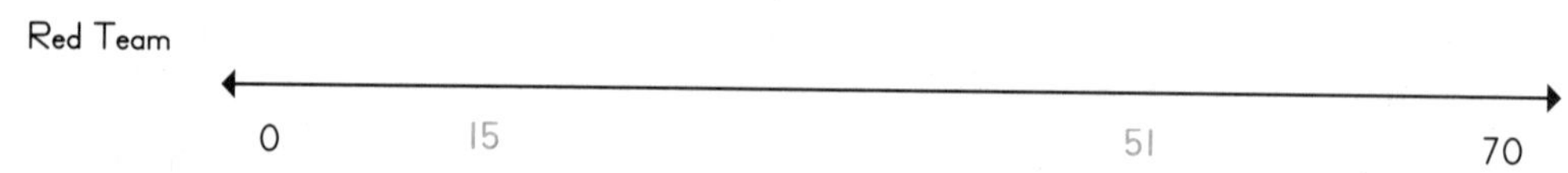

The red team's number line showing the numbers from the first roll.

16. Now do the same for the blue team's number line. Say, "We can put twenty-three about here and then thirty-two about halfway across. Our line is just to help us remember the numbers we are using. Our lines don't have to be perfectly straight or measured."

The blue team's number line showing the numbers from the first roll.

17. Ask students, "How would you determine the difference between the numbers on your team's number line?" Remind students they can jump from the smaller number to the larger number or jump from the larger number to the smaller number.

18. Do a think-pair-share, then record a student's thinking from each team. Check with students to ensure your recording represents their thinking correctly.

Examples of Student Thinking (Red Team: 15 and 51)

Example A

Nala: "You could jump five to twenty, then thirty more to fifty, then one more, and then you are at fifty-one."

Teacher: "So how many jumps is that?"

Nala: "Five and thirty and one is thirty-six."

Teacher: "I'll write that as an equation. Five plus thirty plus one equals 36":

$$5 + 30 + 1 = 36$$

[Using a red marker, record the jumps on the open number line.]

Number line showing Nala's thinking in determining the difference between the red team's numbers 15 and 51.

Math Talk Tip

Think-Pair-Share

When, as teachers, we ask a question to the whole class, often one student responds. However, when we use the think-pair-share strategy, all students get a chance to talk about the question. This can be a chance for students to rehearse with a partner what they may want to say to the whole class, or an opportunity for them to predict what will happen in a safe setting. At first, young students do not know how to choose a discussion buddy for think-pair-share and may need help by being assigned a partner. With experience, students become comfortable with the think-pair-share strategy and appreciate the opportunity to express their ideas before voicing them to the entire class.

Example B

Jason: "You could go to twenty-five. That's ten more. Then go to thirty-five. That's another ten. So that's twenty so far. Then you can go to forty-five. That's another ten. So now you've jumped thirty. From there it's six more 'cause forty-five plus five is fifty and then there's one more."

Teacher: "I noticed you didn't add six but changed it to five plus one."

Jason: "Well, it's easier to add fives or tens. If you need to add a six, you can use a five and then add the one."

Teacher: "That's right. Sometimes students break up numbers to make them easier to add."

$$15 + (10 + 10 + 10 + 5 + 1) = 51$$
$$15 + 36 = 51$$

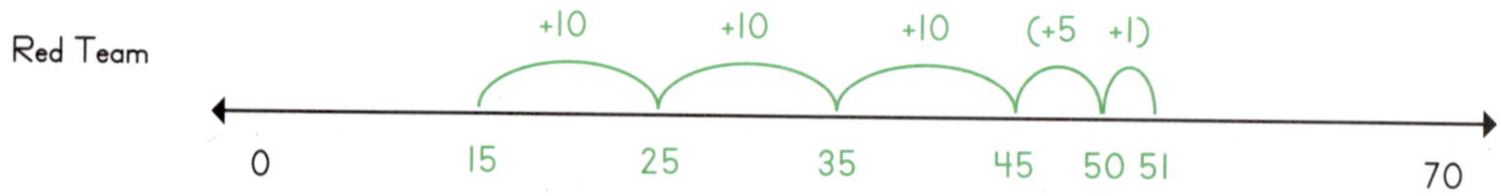

Number line showing Jason's thinking in determining the difference between the red team's numbers of 15 and 51.

Examples of Student Thinking (Blue Team: 23 and 32)

Suejin: "I know that ten more than twenty-three is thirty-three. Then I go back one to get to thirty-two. So the difference is nine."

$$23 + 10 - 1 = 32$$
$$23 + 9 = 32$$

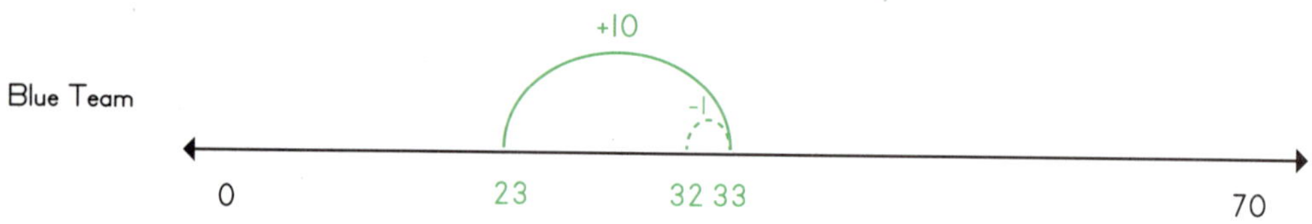

Number line showing Suejin's thinking in determining the difference between the blue team numbers 23 and 32.

19. Enter the differences in the appropriate blanks of the recording sheet. The recording sheet should now look something like this:

> The red team numbers are __51__ and __15__.
> The difference between these numbers is __36__.
>
> 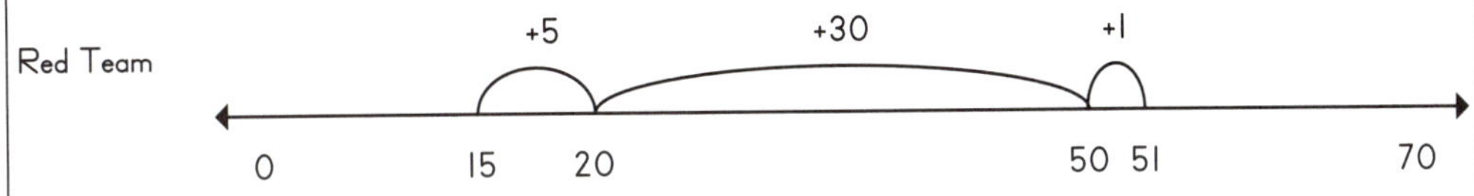
>
>
> The blue team numbers are __23__ and __32__.
> The difference between these numbers is __9__.
>
> 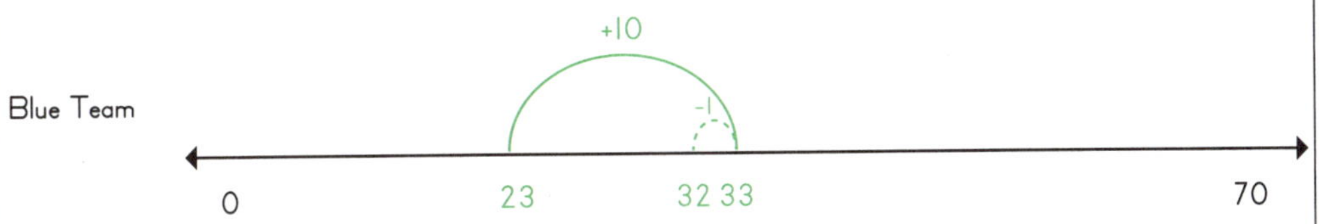
>
>
> The team with the smaller difference is:
> Red Team   Blue Team
>
> (*circle one and record the difference below*)
>
> ______ < ______

20. Now ask the class, "Which team has the smaller difference?" Have students turn and talk.

21. Ask for a volunteer from the team with the smallest difference (in our example, this is the blue team) to come up to the recording sheet and circle the team with the smaller difference.

22. Write the corresponding expression on the recording sheet. In our example, the blue team's difference, nine, is less than the red team's difference, thirty-six, so the expression is 9 < 36.

23. Read the expression aloud. As needed, reference the Greater Than, Less Than, and Equal To Reference Chart (Reproducible 2). Say, "Nine is less than thirty-six. We can also say that a difference of nine is smaller than a difference of thirty-six."

Math Matters!

What Are the Differences?

With two dice labeled *1* through *6*, there are only fifteen possible rolls that result in two different digits.

Remember, in this game, if the roll results in the same two digits, the team must roll again. The rolls are as follows:

1, 2	2, 3	3, 4	4, 5	5, 6
1, 3	2, 4	3, 5	4, 6	
1, 4	2, 5	3, 6		
1, 5	2, 6			
1, 6				

In each case, when the two numbers are formed from the two digits rolled, the differences between the two numbers will be either nine, eighteen, twenty-seven, thirty-six, or forty-five—all multiples of nine.

Math Matters!

The Equal Sign

When red and blue teams compare their differences, they may discover their differences are equal. In this case, they need to write an expression using the equal sign. Some children may have never seen the equal sign used in this way (for example, 18 = 18 or 36 = 36). They may be used to seeing it used only in equations, with two addends on one side and the sum on the right, as in 9 + 18 = 27. Some children think of the equal sign as a kind of "punctuation" that shows the end of a number sentence, instead of thinking about it as a sign that expresses a relationship. Take time to discuss this with students.

Some children may also think that an equation such as 18 = 42 – 24 is not true because there is only one number on the left and two numbers are on the right. With more exposure and explicit conversation about the meaning of the equal sign, students eventually become more comfortable with equations such as 4 + 4 = 5 + 3, in which there are two or more addends on each side of the equal sign.

Teaching Tip

Recording Sheets

Save the students' recording sheets because the sheets give insights on how students are thinking about making moves on a number line.

24. Note: If the two differences are equal, write an equation using the equal sign instead of the less-than sign (see the Math Matters! tip for a description of this situation).
25. Facilitate several more rounds with new volunteers from the red and blue teams until students show confidence in playing the game.

Explore

26. Divide students into groups of four to play the game independently. Provide each group with a red marker and a blue marker, two dice, and a recording sheet (Reproducible 22).
27. Have the groups of four divide into two teams of two (a red team and a blue team) and roll the dice to determine which team goes first (the team with the smaller roll starts). Suggest that students take turns recording with their partner. When one student is recording, the other rolls the dice.
28. Circulate around the room and support students when necessary by asking key questions (see page 248).

Summarize

29. After students have had the opportunity to play the game independently over a period of time (note that the game can be part of a math workshop learning station), bring the class back together for a whole-class discussion. Choose some of the recording sheets to share as a class. Display the recording sheets where everyone can see them.

30. Ask students to look at the recording sheets and see whether other groups used any of the same numbers they used. Ask, "Did you find the difference between the two numbers in the same way as your classmates?"

31. Have two or three students volunteer explain how they jumped from one number to another in a different way than one of their classmates.

32. Summarize by reminding students that the purpose of the game is to find the difference between two numbers (in any way that makes sense) and determine which difference is smaller.

33. Save the recording sheets and post the following number sentences from them on the board:

    21 – 12 = 9

    32 – 23 = 9

    43 – 34 = 9

    54 – 45 = 9

    65 – 56 = 9

34. Ask students to look at these results from their games. Explain, "These were the winning scores for the teams that had the smaller difference of nine." Ask, "What patterns do you see? What do you notice?"

Examples of Student Thinking

Makina: "The two numbers go in order: one, two, three, and like that."

Jenny: "You always get nine in these."

Martin: "The two numbers are the same as the first two numbers, but in a different order."

Teacher: "You are all noticing that there are different ways to find the smaller difference of nine."

Extend Their Learning!

Drawing Two Number Cards 1 through 9

In this extension, teams use number cards numbered *1* through *9* (see Reproducible 1) instead of the two dice. (Reproducible 1 includes cards numbered *1* through *20*. Use cards *1* through *9* only from this reproducible.) Place the cards in a paper lunch sack. Have the red team draw two numbers from the sack and form two, two-digit numbers. Make sure they write the two-digit numbers on their recording sheet. Instruct the red team to place the numbers back in the sack. Then ask the blue team members to draw, form, and record their numbers. This version gives students the opportunity to work with a larger range of numbers, such as ninety-two and twenty-nine, but still have differences that are multiples of nine.

Using Three Dice Instead of Two

In this extension, use three dice instead of two dice to compare two, three-digit numbers. With one roll of three dice, it is possible to create six different three-digit numbers; so, in this version, students use number sense and estimation to decide how to determine the place values for the three digits to obtain the smallest possible difference. For example, if the three digits are 2, 6, and 5, the difference between 256 and 265 would be the smallest difference possible with these numbers. Consider adding the rule that if two of the three dice rolled have the same number, players must roll the three dice again until all three are different numbers.

Examples of Student Thinking

Chelsea (red team): "I rolled a two, a five, and a six."

Teacher: "What two numbers can your team create with these three digits?"

Chelsea: "We could have two hundred sixty-five and five hundred sixty-two."

Teacher: "Are there other numbers you could make that would have a smaller difference?"

Tom: "If we had our numbers in the same hundreds, they would be closer."

Teacher: "It sounds like you have an idea how to make two numbers so they are closer to each other."

Ayumi: "Let's do five hundred twenty-six and five hundred sixty-two." [Everyone nods in agreement.]

Teacher: "It looks like the red team has decided on their numbers. Record your numbers on the recording sheet. Now let's have the blue team roll their three dice and make their numbers."

Brendan (blue team): "I rolled a four, a one, and a six."

Teacher: "What two numbers can your team create with those digits?"

Brendan: "We could have four hundred sixteen and one hundred sixty-four."

Teacher: "Are there other numbers you could make that would have a smaller difference?"

Alya: "One of our numbers is going to be one hundred forty-six. The other is going to be one hundred sixty-four, because then they would be really close to each other."

Teacher: "Record your numbers on the recording sheet. Now, each team needs to determine the difference and share your thinking. Let's start with the red team."

Lizzy (red team): "We jumped to five hundred thirty, and that was four. Then we jumped thirty more to five hundred sixty, and that is thirty-four so far. Then we jumped two to five hundred sixty-two. Our three jumps were thirty-six away."

Teacher (after recording): "Does this look like what you said, Lizzy?"

$$526 + 4 + 30 + 2 = 562$$

$$526 + 36 = 562$$

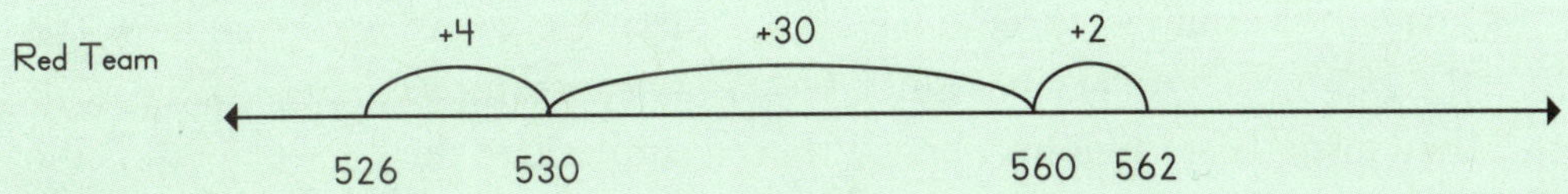

Lizzy: "Yes."

The red team numbers are 526 and 562.

The difference between these numbers is 36.

Teacher: "Will someone from the blue team please let me know how you determined your difference?"

Arman (blue team): "We jumped four to one hundred fifty, then ten to one hundred sixty, then four to one hundred sixty-four. Our difference is eighteen."

Teacher (after recording): "Would you agree that this represents your thinking, Arman?"

$$146 + (4 + 10 + 4) = 164$$

$$146 + 18 = 164$$

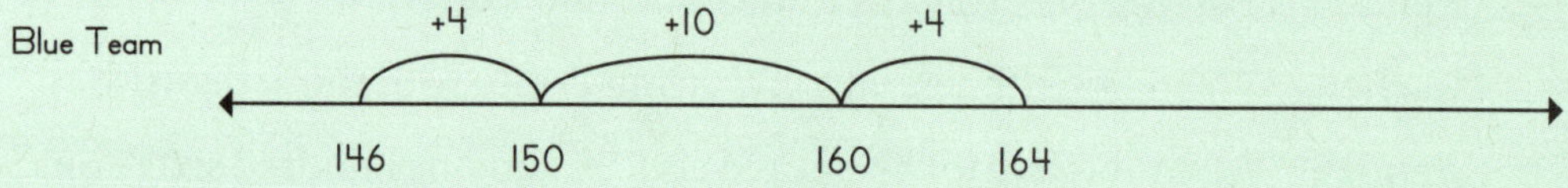

Arman: "Yes."

Teacher: "Which team has the smaller difference?"

The blue team numbers are __146__ and __164__.

The difference between these numbers is __18__.

The team with the smaller difference is: Red Team (Blue Team)
(*circle one and record the difference below*)

__18__ < __36__

To summarize this version of the game, ask students to describe other ways they could determine the differences between their two numbers using different-size jumps. If other combinations of the three digits could have been used to create a smaller difference, prompt students to try those combinations.

Drawing Three Number Cards, 1 through 9

Instead of rolling three dice, place number cards 1 through 9 (see Reproducible 1) in a paper lunch sack and have teams each draw three cards, then form three-digit numbers. Proceed similarly to the previous extension. This game can be challenging for young students to find the least possible difference for their two, three-digit numbers, but the ability to decide which numbers to create from three different digits is surprisingly powerful for some students, and they seem to challenge themselves to think deeply about the size of the numbers they are using.

Adding Nines, Tens, and Elevens

Game 10

Overview

In this game, students take turns recording moves of nine, ten, or eleven along an open number line. Each move is determined by first spinning the game's spinner. The objective is to be the first team to get to the end of the number line.

This game supports students in becoming comfortable adding nine or eleven using their knowledge of adding tens. Begin with having students consider how adding eleven can be thought of as adding ten and then adding one more. Similarly, adding nine can be thought of as adding ten and then moving back one.

Time

45 minutes

Materials

large paper clip, 1 per each pair of students

open number line labeled *0* and *100*, 2 per group of four students

Adding Nines, Tens, and Elevens Spinner (Reproducible 23), 1 per group of four students

red and blue colored markers, 1 of each color per group of four students

Related Games

- G-4 Race to 50
- G-5 The Game of Pig on a Number Line
- G-6 101 and Out!
- G-7 Hot Lava Bridge: Forward and Back, More and Less

Key Questions

- How can you know how far to move without counting by ones?
- How do you know you've landed in the correct place?

Time Saver

Preparing the Open Number Line
Instead of making consumable copies of the number line, laminate the copies or place them in plastic sleeves and provide dry-erase markers during game play. You may also have students draw their own number line; in this case, distribute whiteboards and dry-erase markers.

Teaching Directions

Introduce

1. Prepare a demonstration open number line labeled *0* and *100*. Display it on the wall at student eye level.

An open number line for the game *Adding Nines, Tens, and Elevens.*

A Child's Mind . . .

Counting by Ones

Don't be alarmed if a student suggests starting at 0 and making eleven moves of one to get to 11. Many students go back to what they know when they are faced with something they perceive as "hard." Allow them to finish, then encourage them to use their knowledge of tens and ones to make a move to eleven. Encourage them to trust using tens and ones without needing to count by ones each time.

Also, some children are confused about counting on and may count the number they are on rather than the next number on the number line. These students will likely need to begin with counting one more, then eventually, with more confidence, they can count two more, three more, and so on. Board games that require students to roll dice and move markers are helpful in supporting children to make sense of how to count on from a starting number or place.

Notice how comfortable students are when they add up eleven. If they persist in counting by ones, perhaps they will benefit from looking for patterns in the hundreds chart as well as the number line. If they are still not able to notice patterns in the tens digits, then consider using an open number path (see the Differentiating Your Instruction tip at the end of this game).

2. Gather students on the rug so that everyone has a good view of the number line. Share with them that they will be playing the game *Adding Nines, Tens, and Elevens.* In addition to the open number line, show students the *Adding Nines, Tens, and Elevens* spinner (Reproducible 23).
3. Explain the game by saying, "First we spin this spinner, which tells us whether we can move nine, ten, or eleven spaces along this number line. The goal is to be the first team to get to the end of their number line."
4. Begin with a discussion about how the moves can be made. For example, show the spinner face and point to the section labeled *+11*. Ask, "How can we add eleven without making eleven little jumps?" If students do not suggest jumping ten and then one, say the following, "How could you use a jump of ten in your move to eleven?"
5. Have a volunteer student share their thinking and record it on the open number line. Use a marker to show where 11 might be on the number line and draw the jumps as described.

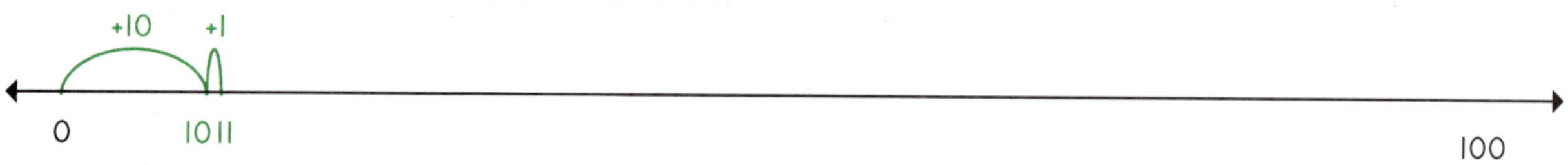

The corresponding move on the number line when the spinner lands on +11.

6. Pretend the spinner keeps landing on +11 and continue asking students to describe how they can add eleven. Use think-pair-share as needed.

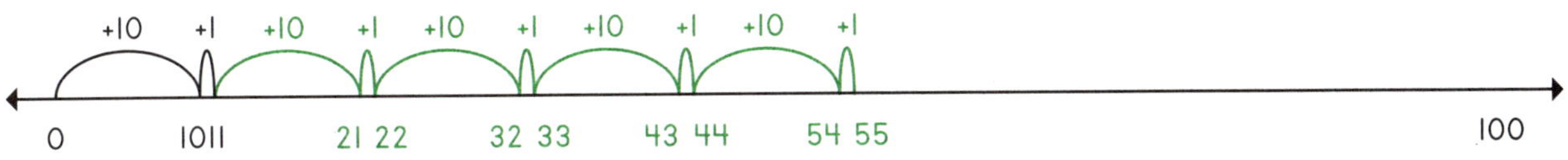

The corresponding moves on the number line after four spins of +11.

7. As students explain their thinking for each +11 move, in addition to recording their jumps on a number line, record the numbers in a vertical chart, which helps make the patterns in the numbers more visible:

Adding Elevens

11
22
33
44
55
66
77
88
99

8. Draw students' attention to the chart and ask, "What do you notice about these numbers?"

> **Math Talk**
>
> **Think-Pair-Share**
>
> When, as teachers, we ask a question to the whole class, often one student responds. However, when we use the think-pair-share strategy, all students get a chance to talk about the question. This can be a chance for students to rehearse with a partner what they may want to say to the whole class, or an opportunity for them to predict what will happen in a safe setting. At first, young students do not know how to choose a discussion buddy for think-pair-share and may need help by being assigned a partner. With experience, students become comfortable with the think-pair-share strategy and appreciate the opportunity to express their ideas before voicing them to the entire class.

Examples of Student Thinking

Ramona: "They're the double numbers."

Abdul: "On the hundreds chart they go down like stair steps."

Hinata: "Each one is one more going down."

9. Ask, "What will the next number be?" Have students think-pair-share. Have a volunteer share their thinking. Does the answer fit the pattern in the chart?
10. Conclude this game, "Because the game is over when we reach one hundred, we are going to stop adding elevens."
11. Now say, "Let's pretend our spinner always lands on plus nine. We're going to practice a few rounds of adding nine." Show the spinner face and point to the +9. Ask, "What move might we make on the number line using what we know about nine as ten minus one?"
12. Record students' thinking on a new number line.

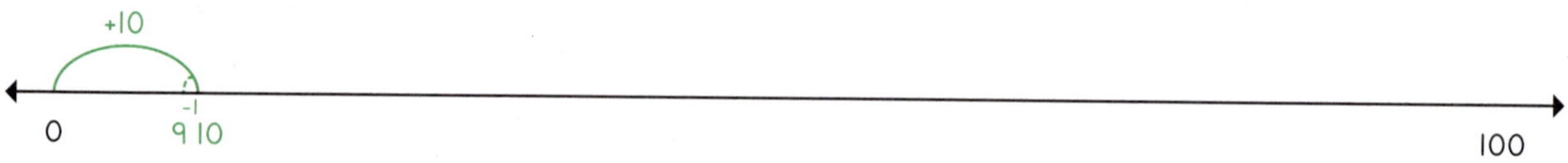

The corresponding move on the number line when the spinner lands on +9.

13. Pretend the spinner keeps landing on +9 and continue asking students to describe how they can add nine. Use think-pair-share as needed.

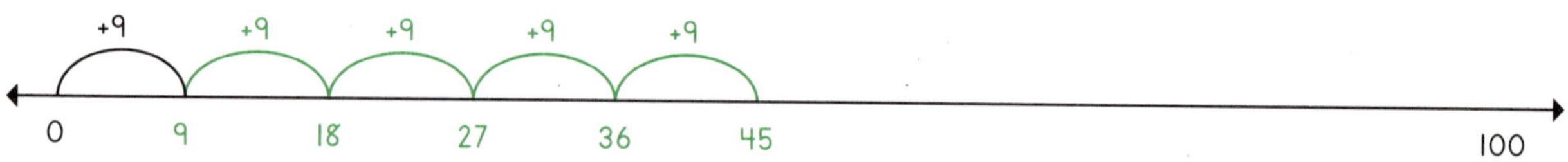

The corresponding moves on the number line when the spinner continues to land on +9.

14. As students explain their thinking for each plus-nine move, in addition to recording their jumps on a number line, record the numbers in a vertical chart, which helps make the patterns in the numbers more visible:

Adding Nines

9
18
27
36
45
54
63
72
81
90
99

15. Draw students' attention to the chart and ask, "What do you notice about these numbers?"

Examples of Student Thinking

Ayumi: "The numbers go one less and one more as you go down."

Teacher: "Could you explain what you mean by going one less and one more?"

Ayumi: "Well, in the ones place it goes down one: nine, eight, seven, six, and like that. In the ten's place it goes up one more: one, two, three, four, up to nine."

16. Ask, "What will the next number be?" Have students think-pair-share. Have a volunteer share their thinking. Does the answer fit the pattern in the chart?

17. Conclude this game, "Because the game is over when we reach one hundred, we are going to stop adding nines."

Assessment Opportunity

Looking for Patterns

When nine or eleven is added continually to other numbers, patterns start to emerge. As you play rounds of this game, such as those shown so far (in which the same number is added each time), keep asking students to describe any patterns they see in the numbers. Some may notice how the ones digit increases or decreases as the numbers go down the column. Ask students to predict which numbers will come next in each column. Here are two more examples, this time starting at 5 and 8 on the number line, instead of 0.

Starting from five . . .

Adding Nines	Adding Elevens
14	16
23	27
32	38
41	49
50	60
59	71
68	82
77	93
86	104
95	

Starting from eight . . .

Adding Nines	Adding Elevens
17	19
26	30
35	41
44	52
53	63
62	74
71	85
80	96
89	107
98	

18. Now that students have had the chance to practice adding both elevens and nines for several moves, it's time to play an actual game. Say, "Now let's play the game as a whole class. There will be two teams: the red team and the blue team." Divide the class so one side is the red team and one side is the blue team. Display an open number line for each team and include a marker with each (ideally a red marker with the red team's number line and a blue marker with the blue team's number line).

19. To determine which team goes first, have a representative from each team spin the spinner. The team with the larger number goes first.

20. A student from the team that goes first spins the spinner and uses their team's colored marker to record the corresponding move on the team's open number line. For example, Arturo, on the red team, spins the spinner and gets a +10. Using the red marker, he shows a jump of ten on his team's number line.

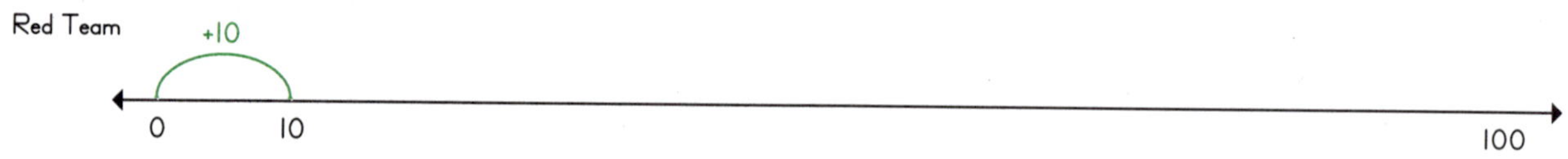

The corresponding move on the red team's number line when the spinner lands on +10.

21. Now a student on the blue team, LaShanna, spins the spinner. The spinner lands on +11. Using the blue marker, LaShanna records the move on her team's number line.

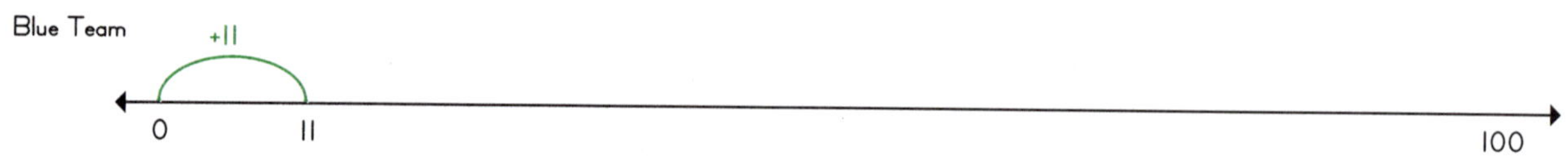

The corresponding move on the blue team's number line when the spinner lands on +11.

Example of Student Thinking

Teacher: "How can you know how far to jump without counting each place?"

LaShanna: "For the first jump, you don't need to count, you just start there."

22. Teams alternate turns until a team's moves takes them past one hundred on their number line.

Examples of Student Thinking

Jin Ae (Red Team): "I spun a nine. I went to nineteen because nineteen is ten and nine."

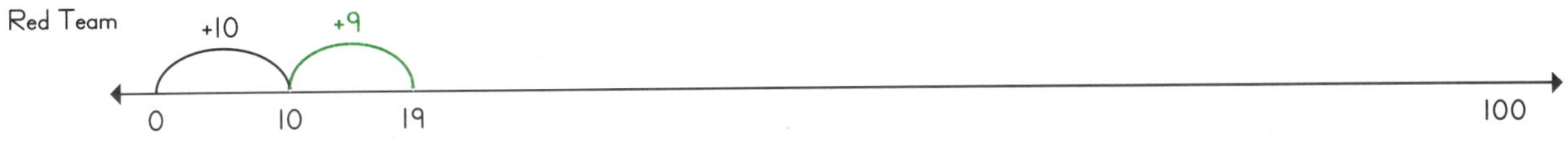

The corresponding second move on the red team's number line when the spinner lands on +9.

Malik (Blue Team): "I spun eleven for the blue team. I went one more first and then ten, but I could've jumped ten to twenty-one and then one more after."

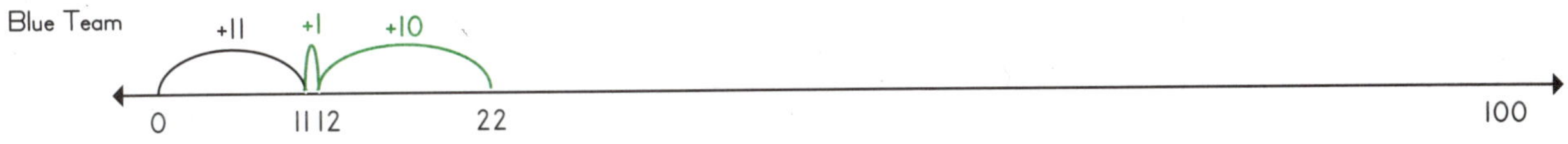

The corresponding second move on the blue team's number line when the spinner lands on +11.

23. As each team sends a representative to take the next turn, ask the rest of the class, "What number will the team land on next if their spin is plus ten?" Then ask, "What about if the spin is plus nine? Plus eleven?" In this way, everyone in the class is involved and thinking about the next possible move.

Math Matters!

Unitizing

Eventually, we want students to think of a group of ten as ten jumps of one step each, as well as one jump of ten. If ten becomes a "friendly" number to them, then nine and eleven become much more than nine ones or eleven ones.

24. The game ends when one team gets to one hundred or beyond one hundred. (Note that the scores are usually very close until the very end.)

Explore

25. After students are comfortable with the game procedures, give students the opportunity to play the game independently in groups of four (so two students are the blue team and the other two are the red team).

26. Provide each group with two open number lines (see the "Teaching Tip", Preparing the Open Number Line earlier in this game), a spinner, and two markers (red and blue).

27. Remind students of the directions. Say, "Take turns spinning the spinner and moving the corresponding number on your own number line. Say your thinking aloud as you record your move. Help your partner by listening to each other's thinking. Check each other's work."

28. Circulate and observe as students play. For example, do some students notice that a jump of eleven followed by a jump of nine is the same as a jump of twenty? Ask key questions (see page 261) as appropriate.

Summarize

29. After students have had the opportunity to play the game independently over a period of time (note that the game can be part of a math workshop learning station), gather them for a whole-class discussion. Ask, "How many turns did you take before you reached one hundred?" (Students will share by counting up the number of jumps on their open number line.) You may also ask, "Did anyone have mostly jumps of nines or elevens?" (Some students may have fewer jumps on their number line because they spun 11 more often.)

Differentiating Your Instruction

Using an Open Number Path

For students who may need additional support, instead of using an open number line, use a red and yellow open number path. Students make their moves by pinning a clothespin or moving a marker along the number path. The use of an open number path encourages students to think about ten as two groups of five. The open number path also helps students begin to notice how to jump ten spaces because the color and location of the squares provide visual clues. For example, notice how the multiples of tens fall on the last yellow square in each group, and how the shaded squares represent the one, two, three, four, and five digits in the ones place.

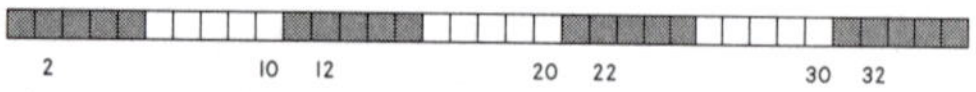

Students can use visual clues of the open number path to see how adding ten to a number like 2 always ends up on the second red square in the pattern.

Students can use these visual clues to add nine or eleven, if they are comfortable with adding ten.

30. Display a sketch of the following spinner face. Say to students, "I'm wondering what would happen in this game if we used this spinner instead?"

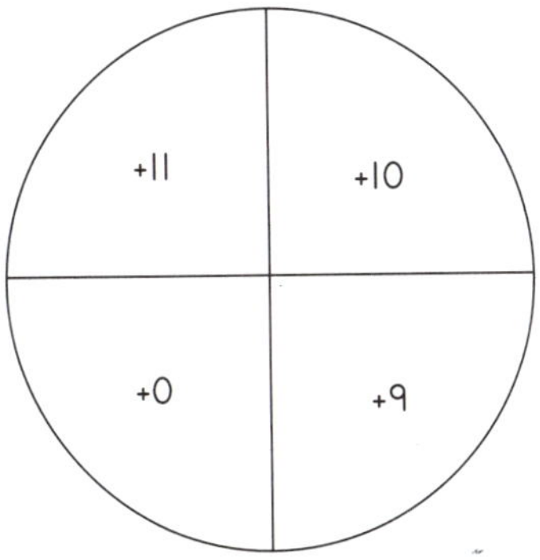

Examples of Student Thinking

Ramona: "It would take longer to finish the game."

Hinata: "When you land on zero you would say, *Oh no, not again!*"

Teacher: "You're right, it might make the game move more slowly. You have been adding nines, tens, and elevens in these games. I hope you found ways to do this now without having to count one by one. Some of you noticed it is easier to just add ten and then take one away or add one more."

Game 11 Get to the Target (Version 1)

Time

20–30 minutes

Materials

large paper clip, 1 per each pair of students

pencil, 1 per student

Get to the Target (Version 1) Spinner (Reproducible 24), 1 per each pair of students

Get to the Target (Version 1) Recording Sheet (Reproducible 25), 1 copy per each pair of students

Sentence Frames Chart (Reproducible 26), 1 copy enlarged and displayed on a wall

Overview

In *Get to the Target* (Version 1) students spin a spinner (labeled *0* through *5*) twice. The two numbers on which the spinner lands form the digits of a target number. After the two-digit target number is created, students try to find ways to get to the target number on an open number line labeled *0* and *60* by jumping forward or backward using only four types of jumps: ones, fives, tens, and twenty-five. This version of *Get to the Target* is most appropriate for first graders.

Related Lessons and Games

- L-4 Jumping by Ones and Tens
- L-5 Jumping by One, Fives, Tens, and Twenty
- G-12 Get to the Target (Version 2)
- G-13 Get to the Target (Version 3)

Key Questions

- How did you keep track of your number of jumps?
- How did you decide whether to start at zero or sixty?
- Which numbers are easy targets to reach? Why?
- Which numbers take more jumps to reach? Why?

Teaching Directions

Introduce

1. Display an enlarged version of the recording sheet (Reproducible 25) so everyone can see it. Draw an open number line labeled *0* and *60* in the "My Number Line and Thinking" space on the recording sheet.

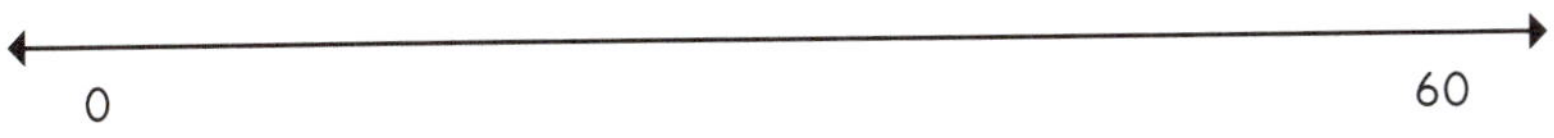

An open number line for the game of *Get to the Target (Version 1)*

Technology Tip

Using an Interactive Whiteboard
If you choose to use an interactive whiteboard, use the interactive pen to draw the open number line on the screen.

2. Explain, "We are going to play a version of the game *Get to the Target.* We'll first play several rounds as a class, and then you will play it with a partner. To start, we must create a two-digit target number. We do this by spinning the spinner twice. The two numbers spun become the digits of our target number. For example, if you spin a four and a three, our target number can either be forty-three or thirty-four. Both digits must be different (otherwise, spin again) and the target number must always be within zero and sixty."
3. After asking two student volunteers to use the spinner (Reproducible 24), write the target number in its approximate place on the open number line. In our example, the target number is 34 so we put it about in the middle of the open number line. Now introduce the objective of the game: Students must find ways to "get to the target" on the number line.
4. Write on the board: *25, 10, 5, 1,* ⇦ or ⇨.
5. Say, "These numbers represent the number of places you can jump along a zero-to-sixty number line during the game. You can only

choose from these numbers when deciding your jumps to get closer to the target number. The arrows remind you that your jumps can be made forward or backward on the number line. Because you can move forward or backward, you can start your jumps at the beginning or the end of the number line."

6. Ask, using our example, "How can we get to our target number of thirty-four? Remember, we can only use these four kinds of jumps: twenty-five, ten, five, and one."
7. Think aloud, "I think I could jump by fives to thirty. Five [counting on fingers], ten, fifteen, twenty, twenty-five, thirty. Then I could do four jumps to get to thirty-four. Yes, that would work."
8. Ask, "Is there another way we could jump to thirty-four? Talk to your partner about a way you could get to the target number, thirty-four, using only the kinds of jumps listed on the board."
9. After students have shared with a partner, ask, "Who would like to share a way you talked about?"

> **Math Talk**
>
> **Discussion Buddies**
>
> Often, students don't know to whom to turn when teachers ask them to talk to a partner. To save time and alleviate any issues that might arise, assign "discussion buddies" before class discussions. Consider keeping the discussion buddies for several weeks so students become comfortable with their partner, then switch buddies so students have opportunities to work with other classmates.

Example of Student Thinking

Farhad: "I started at the beginning and jumped three tens and four ones."

Record Farhad's thinking on the recording sheet's open number line. Explain how to fill out each section on the recording sheet as part of doing this.

"So Farhad started at the beginning, so I am going to circle Starting at Zero. Then I'll draw his jumps like this. I can see you made seven jumps."

Our Target Number: 34

Starting at Zero (circled) or **Starting at 60** (*circle one*)

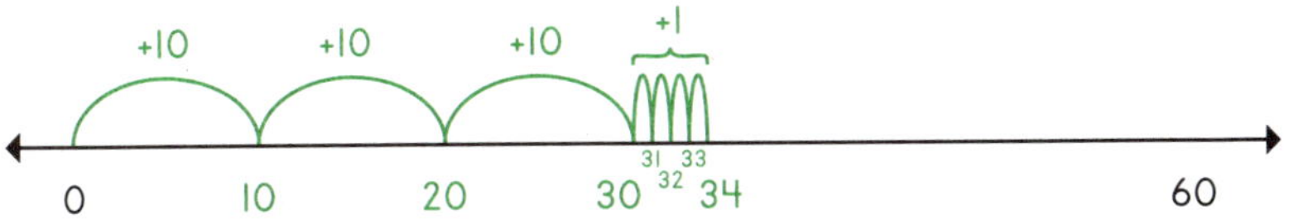

Number line showing Farhad's thinking in getting to the target number 34, starting at 0

10 + 10 + 10 + 1 + 1 + 1 + 1 = 34

Total Number of Jumps: 7

Check with the student to see if the recording represents what he said.

10. Now challenge students to try to get to the target number using less than seven jumps. Remind them they can go backward as well as forward. Use the second half of the recording sheet to record another student's thinking.

Example of Student Thinking

Alicia: "You can go to twenty-five and then jump ten more to thirty-five and then back one to thirty-four and you're there."

Our Target Number: 34

Starting at Zero (circled) or **Starting at 60** (*circle one*)

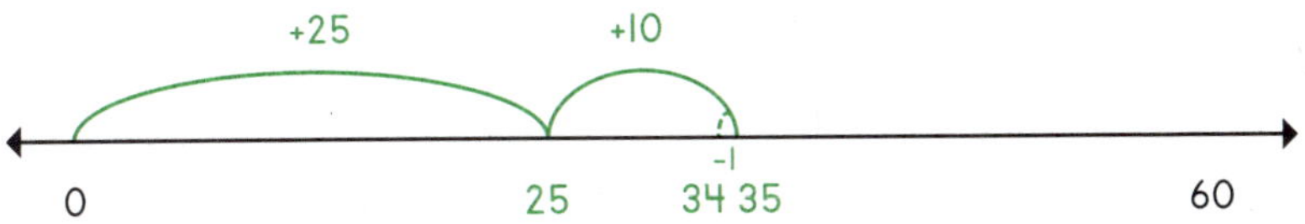

Number line showing Alicia's thinking in getting to the target number 34

25 + 10 − 1 = 34

Total Number of Jumps: 3

A Child's Mind . . .

Making Jumps

At first, students may need to draw jumps on the number line one place at a time. When they begin to realize that jumping by ten or five moves them farther along the number line in fewer jumps, they begin to use those strategies. After they are comfortable with knowing how to move by ones, fives, tens, and twenty-five, they begin to focus on making the fewest jumps when it is their turn. After several turns or games, students may find shortcuts and move even more efficiently on the number line.

Teaching Tip

Starting at Zero or Sixty

If the target number is a large number, encourage students to start at sixty. Ask them why this might be a good choice.

11. Now that two different ways of getting to the target number have been recorded, model how to complete the last line of the recording sheet. Say, "Because Alicia used less jumps than Farhad, we circle Less Jumps Than My Partner."

Less Jumps Than My Partner or **More Jumps Than My Partner** (*circle one*)

12. Play another round of the game. Display a clean copy of the recording sheet and focus on starting from the end of the number line rather than the beginning. Ask, "Is there a way to get to thirty-four starting at sixty instead of zero and still using just the four kinds of jumps: twenty-five, ten, five, and one?"

13. Say, "Talk with your partner and see if you can describe the jumps."

Example of Student Thinking

Avik: "You could go back twenty-five to thirty-five and then go back one more to thirty-four."

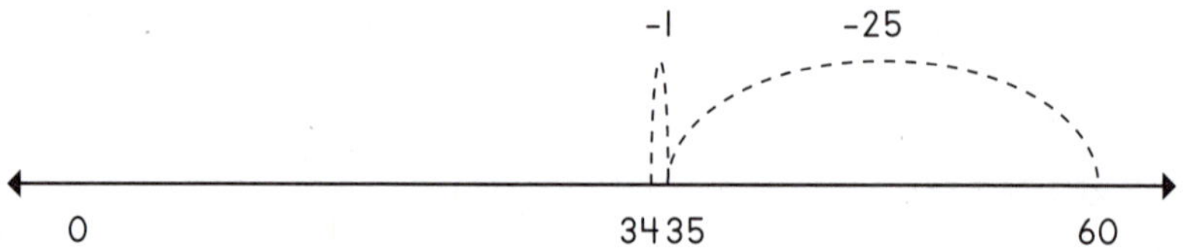

Number line showing Avik's thinking in getting to the target number, 34, starting at 60.

$$60 - 25 - 1 = 34$$

14. Play another round of the game with a new two-digit target number, this time doing *less* thinking aloud. Instead, let students come up to the recording sheet and describe their jumps on the number line. To help students in articulating their thinking, display the following sentence frames (Reproducible 26):

    First, I started at ______.

    Then I jumped ______ to ________.

    Next, I jumped ______ more (or less) to ______.

    Then I jumped ______ more (or less) to ______.

    Finally, I jumped ______ to the target number of ______.

Explore

15. When students are comfortable with the game procedures, have them play the game independently with a partner. Provide each pair with a spinner, two pencils, and a copy of the recording sheet (Reproducible 25).

16. Have partners determine who is Partner A and who is Partner B. Partner A should fill out the top half of the recording sheet, and Partner B, the bottom half.

17. Have partners create the target number together (spinning the spinner twice) and record it on the recording sheet.

18. Then, instruct them to draw their number lines and record their thinking in the appropriate sections on the recording sheet.

19. Remind them to look at each other's number lines and determine who made the fewest jumps. Have them circle their observation (Less Jumps Than My Partner or More Jumps Than My Partner) on the recording sheet.

20. Circulate around the room and support students when necessary by asking key questions (see page 270).

Teaching Tip

Recording Sheets

Save students' recording sheets because the sheets give insights on how students are thinking about making moves on a number line. They can be used as an assessment tool.

Summarize

21. After students have had the opportunity to play the game independently over a period of time (note that the game can be part of a math workshop learning station), gather them together for a whole-class discussion.

22. Choose some of the recording sheets to share as a class. Display the recording sheets where everyone can see them.

23. Ask students to look at the recording sheets and see whether other groups were able to get to their target with only a few moves. Ask, "Did you find that some students were able to use less moves to a target number than you did?"

24. Have two or three students volunteer how they jumped from 0 or 60 to the same target number.

25. Ask, "Which target numbers are easy to jump to? Why?"

26. Ask, "When should you start at zero?" "When should you start at 60? Why?"

Examples of Student Thinking

Bae: "When the target number is twenty-five or ten, they are easy because you only have to do one jump"

Florence: "I think it's easier to jump up from zero than to go back from sixty because you are adding up instead of subtracting back."

Mihir: "Yes, but when the number is really close to 60 like 55, it's easy to go back."

27. Tell students that there are other versions of this game that they can play in the future that use larger jumps and there will be larger numbers on the number line. Let them know that you will leave the copies of the recording sheets on display so they can look them over and see some of the different strategies that their classmates used.

Get to the Target (Version 2)

Game 12

Overview

In *Get to the Target* (Version 2), students spin a spinner (labeled *0* through *9*) twice. The two numbers on which the spinner lands form the digits of a target number. After the two-digit target number is created, students try to find ways to get to the target number on an open number line labeled *0* and *100* by jumping forward or backward using only four types of jumps: fifty, tens, fives, and/or ones. This version of *Get to the Target* is most appropriate for first and second graders.

Related Lessons and Games

- L-4 Jumping by Ones and Tens
- L-5 Jumping by One, Fives, Tens, and Twenty
- G-11 Get to the Target (Version 1)
- G-13 Get to the Target (Version 3)

Key Questions

- How did you keep track of your number of jumps?
- How do you know whether to start at zero or one hundred?
- Which numbers are easy targets to reach? Why?
- Which numbers take more jumps to reach? Why?

Time

20–30 minutes

Materials

large paper clip, 1 per each pair of students

pencil, 1 per student

Get to the Target (Version 2) Spinner (Reproducible 27), 1 per each pair of students

Get to the Target (Version 2) Recording Sheet (Reproducible 28), 1 copy per each pair of students

Sentence Frames Chart (Reproducible 26), 1 copy enlarged and displayed on a wall

Technology Tip

Using an Interactive Whiteboard
If you choose to use an interactive whiteboard, use the interactive pen to draw the open number line on the screen.

Teaching Directions

Introduce

1. Display an enlarged version of the recording sheet (Reproducible 28) so everyone can see it. Draw an open number line labeled *0* and *100* in the "My Number Line and Thinking" space on the top half of the recording sheet.

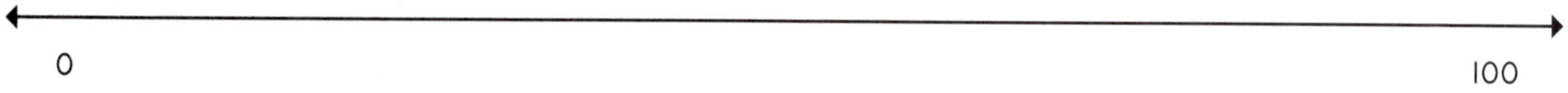

An open number line for the game of *Get to the Target (Version 2).*

2. Explain, "We are going to play a version of the game *Get to the Target.* We'll first play several rounds as a class, and then you'll play it independently with a partner. To start, we must create a two-digit target number. We do this by spinning the spinner twice. The two numbers spun become the digits of our target number. For example, if you spin a five and a seven, our target number can either be fifty-seven or seventy-five. Both digits must be different (otherwise, spin again) and the target number must always be within zero and one hundred."
3. After asking two student volunteers to use the spinner (Reproducible 27), write the target number in its approximate place on the open number line. In our example, the decided target number is 57. Now introduce the objective of the game: Students must find ways to "get to the target" on the number line.
4. Write on the board: *50, 10, 5, 1,* ⇐ or ⇒.
5. Say, "These numbers represent the number of places you can jump along a zero-to-one hundred number line during the game. You can only choose from these numbers when deciding your jump to get closer to the target number. The arrows remind you that your jumps can be made forward or

backward on the number line. Because you can move forward or backward, you can start your jumps at the beginning or end of the number line."

6. Using our example, ask, "How can we get to our target number of fifty-seven? Remember, we can only use these four kinds of jumps: fifty, ten, five, and one."
7. Think aloud, "I think I could jump by tens to fifty. Ten [counting on fingers], twenty, thirty, forty, fifty. Then I could do one jump to five and two jumps of one to get to fifty-seven. Yes, that would work."
8. Record your thinking on the top half of the recording sheet. Explain how to fill out each section of the recording sheet as you do. "I started at zero and jumped to fifty, so I'll circle starting at zero.

Our Target Number: 57

Starting at Zero or **Starting at 100** (*circle one*)

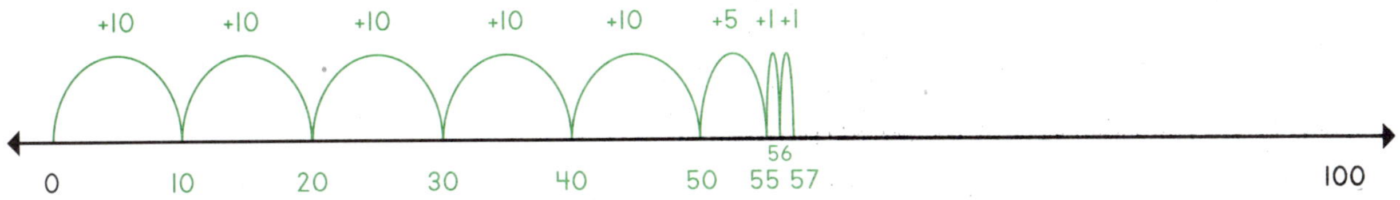

Number line showing the teacher's thinking in getting to the target number 57, starting at 0.

10 + 10 + 10 + 10 + 10 + 5 + 1 + 1

Total Number of Jumps: 8

9. Ask, "Is there another way I could jump to fifty-seven? Talk to your partner about a way you could get to the target number, fifty-seven, using only the kinds of jumps listed on the board."

Math Talk

Discussion Buddies

Often, students don't know to whom to turn when teachers ask them to talk to a partner. To save time and alleviate any issues that might arise, assign "discussion buddies" before class discussions. Consider keeping the discussion buddies for several weeks so students become comfortable with their partner, then switch buddies so students have opportunities to work with other classmates.

10. After students have shared with a partner, ask, "Who would like to share a way you talked about?" In our example, a student might say, "We started at the beginning and jumped fifty, and then we jumped five and then two ones."

11. Record students' thinking using the bottom half of the recording sheet. Explain how to fill out each section on the recording sheet as part of doing this.

Our Target Number: __57__

Starting at Zero (circled) or **Starting at 100** (*circle one*)

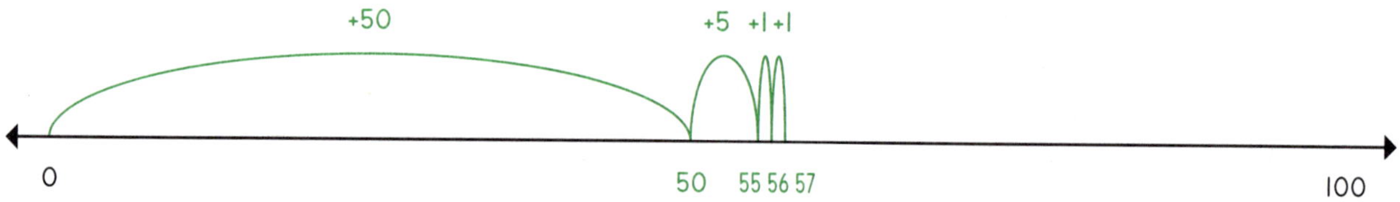

Number line showing a student's thinking in getting to the target number 57, starting at 0.

$50 + 5 + 1 + 1 = 57$

Total Number of Jumps: __4__

A Child's Mind . . .

Making Jumps

At first, students may need to draw jumps on the number line one step at a time. When they begin to realize that jumping by ten or five moves them farther along the number line in fewer jumps, they begin to use those strategies. After they are comfortable with knowing how to move by ones, fives, tens, and fifty, they focus more on making the fewest jumps when it is their turn. After several turns or games, students may find shortcuts and move even more efficiently on the number line.

12. Compare the student's way to the way you thought about getting to the target number. Say, "How does this way compare to the way I thought?"

13. Model how to complete the final line of the recording sheet. Say, "Because you used less jumps than I did, circle More Jumps Than My Partner on the recording sheet."

Less Jumps Than My Partner or

More Jumps Than My Partner (circled) (*circle one*)

14. Play another round of the game. Display a clean copy of the recording sheet and focus on starting from the end of the number line rather than the beginning. Ask, "Is there a way to get to fifty-seven starting at one hundred instead of zero, but still selecting from just the four jumps: fifty, ten, five, and one?"

15. Say, "Talk with your partner and see if you can describe the jumps."

Example of Student Thinking

Mona: "You could go back ten jumps four times to get to sixty and then go back three ones to get to fifty-seven."

Our Target Number: ___57___

Starting at Zero or **Starting at 100** (circled) (*circle one*)

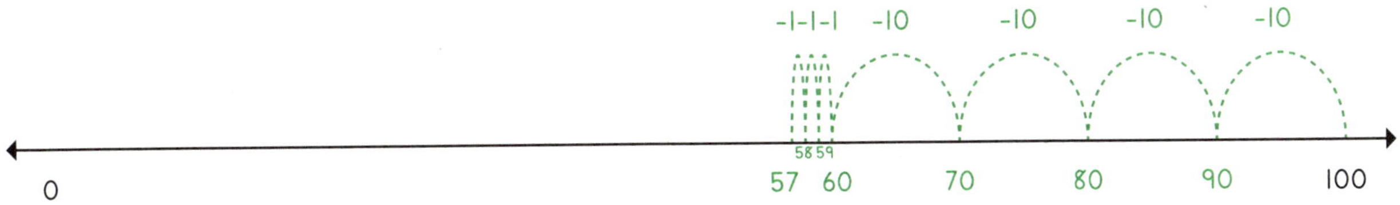

Number line showing Mona's thinking in getting to the target number 57, starting at 100.

100 − 10 − 10 − 10 − 10 − 1 − 1 − 1 = 57

Total Number of Jumps: ___7___

16. Play another round of the game with a new two-digit target number. This time, do less thinking aloud and let students come up to the recording sheet and describe their jumps on the number line. To help students articulate their thinking, display the following sentence frames (Reproducible 26):

    First, I started at ______.

    Then I jumped ______ to ________.

    Next, I jumped ______ more (or less) to ______.

    Then I jumped ______ more (or less) to ______.

    Finally, I jumped ______ to the target number of ______.

Teaching Tip

Starting at Zero or One Hundred?

If the target number is a large number, encourage students to start at one hundred. Ask them why this might be a good choice.

Explore

17. When students are comfortable with the game procedures, have them play the game independently with a partner. Provide each pair with a spinner, two pencils, and a copy of the recording sheet (Reproducible 28).

18. Have partners determine who is Partner A and who is Partner B. Partner A should fill out the top half of the recording sheet, and Partner B, the bottom half.

19. Have partners create the target number together (spinning the spinner twice) and record it on the recording sheet.

20. Then, instruct them to draw their number lines and record their thinking in the appropriate sections on the recording sheet.

21. Remind them to look at each other's number lines and determine who made the fewest jumps. Have them circle their observation (Less Jumps Than My Partner or More Jumps Than My Partner) on the recording sheet.

22. Circulate around the room and support students when necessary by asking key questions (see page 277).

Teaching Tip

Recording Sheets
Save students' recording sheets because the sheets give insights on how students are thinking about making moves on a number line. They can be used as an assessment tool.

Summarize

23. After students have had the opportunity to play the game independently over a period of time (note that the game can be part of a math workshop learning station), gather them together for a whole-class discussion. Focus on linking their thinking to math symbols. Display a copy of a recording sheet as follows.

Our Target Number: 79

Starting at Zero or **Starting at 100** (circled) (*circle one*)

My Number Line and Thinking

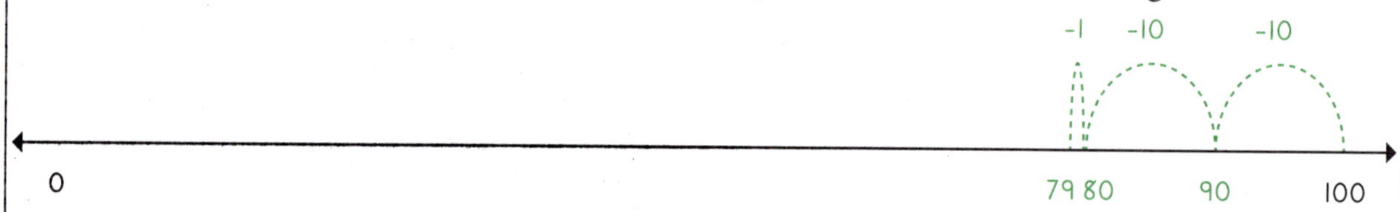

Total Number of Jumps: 3

24. Explain, "We can record the moves shown on the recording sheet using an equation. We know that our target number is seventy-nine. Let's say we started at one hundred and subtracted two groups of ten. So, we could write this:

    $100 - 20 = 80$

    Then we subtracted one from eighty.

    $80 - 1 = 79$

    Putting it all together, the equation is $100 - 20 - 1 = 79$ or $100 - 21 = 79$."

25. Ask key questions to encourage student thinking.

Examples of Student Thinking and Key Questions

Teacher: "Why was it a good idea, in this example, to start at one hundred instead of zero?"

Bae: "You would have to jump all the way to seventy or eighty to get close to seventy-nine."

LaShanna: "Seventy-nine is lots closer to one hundred."

Camden: "We did it in only three jumps when we started at the end."

Serena: "But subtracting is harder to do than addition."

Teacher: "Is it possible to get to seventy-nine from zero? How?"

Adil: "You could jump by fifty and that would be pretty far."

Teacher: "Suppose you could also make jumps of twenty, thirty, forty, fifty, sixty, seventy, eighty, or ninety in this game. How would the game change?"

Kiko: "It would make the game easier. You wouldn't have to think too hard. You just choose the number in the tens and then the ones."

Math Matters!

Flexibility with Computation/Seeing Relationships

When asked how they would teach subtraction of two-digit numbers when "borrowing" or "regrouping" is needed, researchers dealing with how teachers think about teaching mathematics discovered that U.S. teachers and Chinese teachers seem to view this topic differently. U.S. teachers could perform the calculations correctly and could explain how to do them, that is, describe the correct procedure. However, fewer than 20 percent of the U.S. teachers had a conceptual grasp of the regrouping process—decomposing one 10 into 10 ones. By contrast, the Chinese teachers overwhelmingly (86%) understood and could explain the idea of decomposing numbers into more meaningful (or "friendly") chunks. Chinese teachers were more interested in operations and relationships. In particular, they were interested in faster and easier ways to perform a given computation, how the meanings of the four operations are connected, and how the meaning and the relationships of the operations are represented (Liping Ma, *Knowing and Teaching Elementary Mathematics*, Lawrence Erlbaum Associates, 1999).

Teachers in the schools where I work and supervise have found the work in this book with number lines to be useful for thinking about the meaning of number relationships. When we look at the relationships between addition and subtraction, we can solve a subtraction problem by thinking about *adding on* from a known number to the total in order to determine the difference between the two numbers. Thinking about adding on is much easier for children to do than to count backward or subtract.

26. Now encourage students to make connections by asking them to compare the equation 100 – 21 = 79 with the following common error:

$$\begin{array}{r} 100 \\ -\ 21 \\ \hline 81 \end{array}$$

27. Ask, "What did this student do to get the incorrect answer of eighty-one?"

    Some students will recognize that instead of subtracting one from zero in the one's place ("because you can't take one from zero"), the student subtracted zero from one and wrote down one.

A Child's Mind . . .

Constant Differences

Do not expect that this one example will make sense to all students. Over time, when students have many opportunities to work with number lines and number paths, they will see that the distance between two numbers remains the same when both of the numbers are increased or decreased by the same amount. (I was always two years older than my brother when we were children and now as adults I'm still two years older than him!)

28. Now, ask students to consider the equation 99 – 20 = ?

$$\begin{array}{r} 99 \\ -\ 20 \\ \hline \end{array}$$

29. Ask, "Will the distance between these two numbers be the same as the distance between one hundred and twenty-one? How do you know?" Show both on a number line to support your thinking.

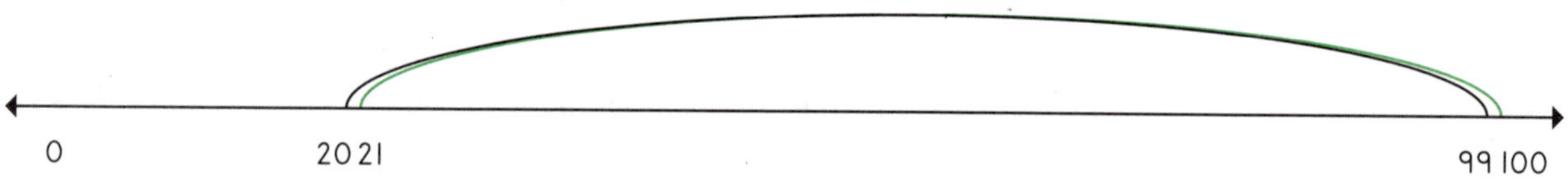

Model of constant differences.

30. Explain that both curves are the same length: jumps of seventy-nine. Say, "So sometimes we can look at the relationships of two numbers to one another and think about how adding one or subtracting one to both numbers does not change the relationship."

Get to the Target (Version 3)

Game 13

Overview

In *Get to the Target* (Version 3), students use a spinner (labeled *0* through *9*) three times. The three numbers on which the spinner lands form the digits of a target number. After the three-digit target number is created, students try to find ways to get to the target number on an open number line labeled *0* and *1,000* by jumping forward or backward using only six types of jumps: five hundred, one hundred, fifty, ten, five, and one. This version of *Get to the Target* is most appropriate for second graders.

Related Lessons and Games

- L-4 Jumping by Ones and Tens
- L-5 Jumping by One, Fives, Tens, and Twenty
- G-11 Get to the Target (Version 1)
- G-12 Get to the Target (Version 2)

Key Questions

- How did you keep track of your number of jumps?
- How do you know whether to start at zero or one thousand?
- Which numbers are easy targets to reach? Why?
- Which numbers take more jumps to reach? Why?

Time

20–30 minutes

Materials

large paper clip, 1 per each pair of students

pencil, 1 per student

Get to the Target (Version 3) Spinner (Reproducible 29), 1 per each pair of students

Get to the Target (Version 3) Recording Sheet (Reproducible 30), 1 copy per each pair of students

Sentence Frames Chart (Reproducible 26), 1 copy enlarged and displayed on a wall

Teaching Directions

Introduce

1. Display an enlarged version of the recording sheet (Reproducible 30) so everyone can see it. Draw an open number line labeled *0* and *1,000* in the "My Number Line and Thinking" space on the top half of the recording sheet.

An open number line for the game of *Get to the Target* (Version 3).

2. Explain, "We are going to play a version of the game *Get to the Target.* We'll first play several rounds as a class, then you'll play it independently with a partner. To start, we must create a three-digit target number. We do this by spinning the spinner three times. The three numbers spun are the digits of our target number. For example, if you spin a five, a nine, and a seven, the target number could be five hundred seventy-nine, five hundred ninety-seven, seven hundred fifty-nine, seven hundred ninety-five, nine hundred fifty-seven, or nine hundred seventy-five. All three digits must be different (otherwise, spin again) and the target number must always be between zero and one thousand."

3. Write the target number in its approximate place on the open number line. In our example, the decided target number is 975. Now introduce the objective of the game: Students must find ways to "get to the target" on the number line.

4. Write on the board: *500, 100, 50, 10, 5, 1,* ⇐ or ⇒.

5. Say, "These numbers represent the size of the jumps you can use along a zero-to-one thousand number line during the game. You can only choose from these numbers when deciding your jump to get closer to the target number. The arrows remind you that your jumps can be made forward or backward on the number line. Because you can move forward or backward, you can start your jumps at the beginning or end of the number line."

6. Using the example, ask, "How can we get to our target number of nine hundred seventy-five? Remember, we can only use six kinds of jumps: five hundred, one hundred, fifty, ten, five, and one."

7. Think aloud, "I think I could jump by hundreds to nine hundred. Then I could do one jump of fifty and two jumps of ten to get to nine hundred seventy. Then I'd only have one jump of five left to get to the target number of nine hundred seventy-five. Yes, that would work."

8. Record your thinking on the top half of the recording sheet. Explain how to fill out each section of the recording sheet as you do. "I started at zero so I'll circle that on my sheet. Then I can draw nine jumps of one hundred, one jump of fifty, two jumps of ten and one jump of five."

Our Target Number: 975

Starting at Zero (circled) or **Starting at 1,000** (*circle one*)

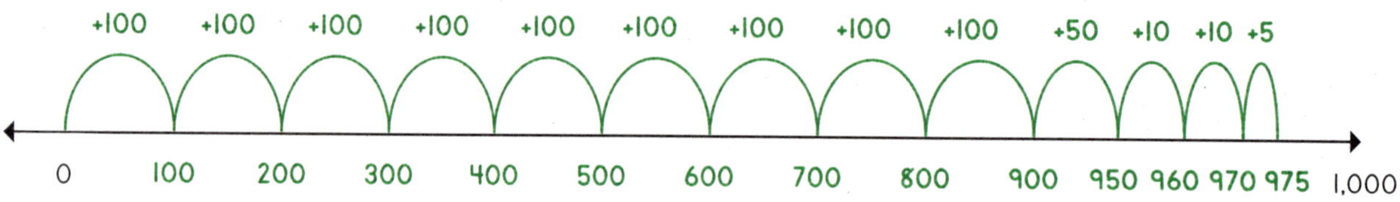

Number line showing a teacher's thinking in getting to the target number 975, starting at 0.

100 + 100 + 100 + 100 + 100 + 100 + 100 + 100 + 100 + 50 + 10 + 10 + 5 = 975

Total Number of Jumps: 13

9. Ask, "Is there another way you could jump to nine hundred seventy-five? Talk to your partner about a way you could get to the target number, nine hundred seventy-five, using only the kinds of jumps listed on the board. See if you can do it in less than thirteen jumps."

10. After students have shared with a partner, ask, "Who would like to share a way you talked about?" In our example, a student might say, "We started at the beginning just like you did, but we jumped five hundred at first and then jumped four hundred more. Then we did the rest the same. We got there with nine jumps."

Math Talk

Discussion Buddies

Often, students don't know to whom to turn when teachers ask them to talk to a partner. To save time and alleviate any issues that might arise, assign "discussion buddies" before class discussions. Consider keeping the discussion buddies for several weeks so students become comfortable with their partner, then switch buddies so students have opportunities to work with other classmates.

11. Record students' thinking using the bottom half of the recording sheet. Explain how to fill out each section on the recording sheet as part of doing this.

Our Target Number: 975

Starting at Zero (circled) or **Starting at 1,000** (*circle one*)

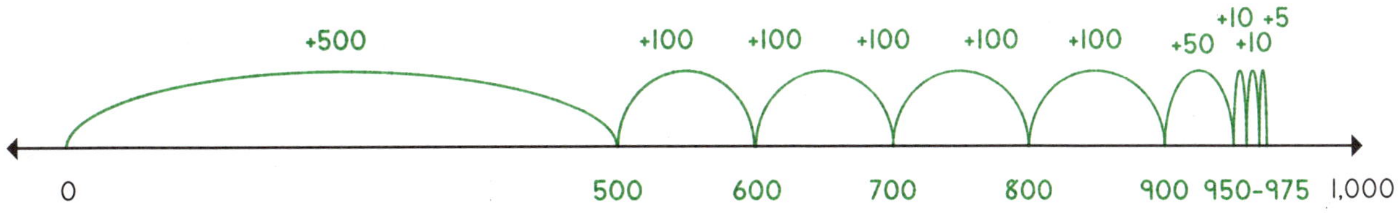

Number line showing a student's thinking in getting to the target number 975, starting at 0.

500 + 100 + 100 + 100 + 100 + 50 + 10 + 10 + 5 = 975

Total Number of Jumps: 9

12. Compare the student's thinking to the way you thought about getting to the target number. Say, "How does this way compare to the way I thought?"

13. A response might be, "His had less jumps." Use this information to complete the final line of the recording sheet. Say, "Yes, you took less jumps than I did, so now we circle Less Jumps Than My Partner."

Less Jumps Than My Partner (circled) or

More Jumps Than My Partner (*circle one*)

14. Play another round of the game. Display a clean copy of the recording sheet and focus on starting from the end of the number line rather than the beginning. Ask, "Is there a way to get to nine hundred seventy-five starting at one thousand instead of zero, still selecting from just the six jumps: five hundred, one hundred, fifty, ten, five, and one?"

15. Say, "Talk with your partner and see if you can describe the jumps."

Example of Student Thinking

Habib: "You could go back two jumps of ten and one jump of five. It's like you're getting to nine dollars and seventy-five cents."

Teacher: "Could you explain how you are thinking about the money?"

Habib: "Well, I thought that one thousand was like ten dollars and no cents, so going backward twenty-five cents gets you to nine dollars and seventy-five cents."

Teacher: "Let's draw Habib's jumps on our one-to-one thousand number line."

Our Target Number: 975

Starting at Zero or **Starting at 1,000** (circled)
(*circle one*)

1,000 − 10 − 10 − 5 = 975

Total Number of Jumps: 3

16. Play another round of the game with a new three-digit target number, this time doing less thinking aloud. Let students come up to the recording sheet and describe their jumps on the number line. To help students articulate their thinking, display the following sentence frames (these are also available as Reproducible 26):

    First, I started at ______.

    Then I jumped ______ to ________.

    Next, I jumped ______ more (or less) to ______.

    Then I jumped ______ more (or less) to ______.

    Finally, I jumped ______ to the target number of ______.

Explore

17. When students are comfortable with the game procedures, have them play the game independently with a partner. Provide each pair with a spinner, two pencils, and a copy of the recording sheet (Reproducible 30).

18. Have partners determine who is Partner A and who is Partner B. Partner A should fill out the top half of the recording sheet, and Partner B, the bottom half.

19. Have partners create the target number together (spinning the spinner three times) and record it on the recording sheet.

20. Then, instruct them to draw their number lines and record their thinking in the appropriate sections on the recording sheet.

21. Remind them to look at each other's number lines and determine who made the fewest jumps. Have them circle their observation (Less Jumps Than My Partner or More Jumps Than My Partner) on the recording sheet.

22. Circulate around the room and support students when necessary by asking key questions (see page 285).

Summarize

23. After students have had the opportunity to play the game independently over a period of time (note that the game can be part of a math workshop learning station), gather them back together for a whole-class discussion. Display some student recording sheets and ask students to notice which target numbers were reached in very few jumps. Ask: "Why do you think some numbers are easier to reach than others?"

Teaching Tip

Recording Sheets

Save students' recording sheets because the sheets give insights on how students are thinking about making moves on a number line. They can be used as an assessment tool.

Teaching Tip

Using Student Strategies

I love it when students think about strategies for making the game more difficult, more fun or better. I try to use their suggestions when they are proposed to honor their thinking and give them some status for their ideas.

Example of Student Thinking

Jiang: "It depends on how careful you are to choose your target number. You can make it easy or hard."

Teacher: "What do you mean by easy or hard? Can you give us an example?"

Jiang: "Well, if you choose a number that is close to the beginning or the end, then that will be easier because it won't take too many jumps to get there."

Teacher: "Is there another way to think about making an easy target number besides how close to the end of the number line it is?"

Ilia: "If you have a target number that is the same as one of the jumping numbers, or one more or one less, then it's easy."

Teacher: "Give us an example of what you mean."

Ilia: "Five hundred ten is five hundred and ten, so that number only takes two jumps. Same as with one hundred fifty."

Teacher: "What if you wanted to make the target number have the most number of jumps instead of the least? What would you do?"

Kwanza: "You could always jump by ones."

Teacher: "What do you think that kind of game would be like?"

Jaslene: "It would take a long time and it wouldn't be any fun."

Race to 1,000

Game 14

Overview

The fast-moving game of *Race to 1,000* provides engaging opportunities for students to practice adding and subtracting single-digit numbers, multiples of ten, and multiples of one hundred. Teams alternate taking turns, first drawing an action card to determine their move on an open number line (adding or subtracting ten, twenty, thirty, forty, fifty, one hundred, two hundred, or three hundred). After the action card has been drawn, the team draws a question card, prompting team members to think more about the number relationships and to practice adding or subtracting.

Because this game is played on an open number line, students have a chance to show jumps in ways that match their level of understanding. For example, if the action card is +100, some students may choose to move by ten jumps of ten whereas others will move one jump of one hundred.

Race to 1,000 is most appropriate for second graders; for kindergartners and first graders, try the games *Race to 50,* which includes *Race to 100.*

Related Game

- G-4 Race to 50

Time

30 minutes

Materials

Count to a Million by Jerry Pallotta (or another favorite counting book)

2 open number lines labeled *0* and *1,000*, one marked *Red Team* and one marked *Blue Team,* 1 set per each red team versus blue team

red marker, 1 per each red team

blue marker, 1 per each blue team

Race to 1,000 Action Cards (Reproducible 31), 1 set per each red team versus blue team (made from 1 copy of the reproducible)

Race to 1,000 Question Cards (Reproducible 32), 1 set per each red team versus blue team (made from 1 copy of the reproducible)

paper lunch sacks, 1 labeled *Action Cards* and 1 labeled *Question Cards,* 1 set per each red team versus blue team (to hold the cards)

Race to 1,000 Recording Sheet (Reproducible 33), 1 copy per student

Time Savers

Preparing the Number Lines

Instead of making consumable copies of the number lines, laminate the copies or place them in plastic sleeves and provide dry-erase markers (red and blue) during game play. You may also have students draw their own number lines; in this case, distribute whiteboards and dry-erase markers.

Preparing the Cards

For the purpose of this game, a set of cards is made with one copy of the reproducible. Copy the action cards and question cards on different-color paper so they can be sorted easily. Shuffle the cards and place them into paper lunch sacks (or in two piles—an action cards pile and a question cards pile).

Key Questions

- How did you know how far to jump on the number line?
- How did you solve what was asked of you on the question card?
- Which action card would you like to draw next? Why?
- Which action card is not a very helpful card to draw? Why?

Teaching Directions

Introduce

1. Prepare a demonstration set of open number lines per the "Materials" list, making sure to label one "Red Team" and one "Blue Team." Display it on the wall at student eye level.

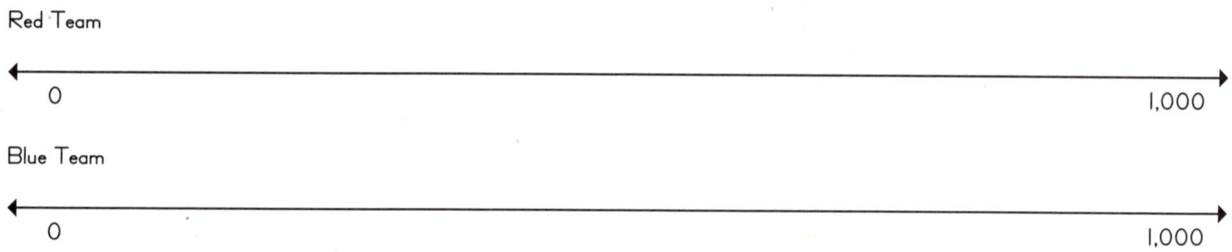

The two open number lines to use when playing *Race to 1,000.*

2. Prepare a demonstration set of both the action cards and questions cards per the "Materials" list. For purposes of modeling the game, place each set in a pile. (After a few rounds of modeling the game, the cards can then be placed in paper lunch sacks.) Also, have a red marker and a blue marker on hand.
3. Gather students on the rug so that everyone has a good view of the number lines and card piles. Seat them close together to make listening to each other easier.

4. If you have the book *Count to a Million* by Jerry Pallotta, set the stage for counting larger numbers by reading it together. (If you do not have this book, choose one from your collection that supports counting larger numbers.)

5. To warm up students even more for counting larger numbers, do a choral count with them, counting by hundreds to one thousand: 100, 200, 300, 400, 500, 600, 700, 800, 900, 1,000. As students count, write the following on the board in a vertical column:

   Counting by Hundreds, Starting with 100

   100
   200
   300
   400
   500
   600
   700
   800
   900
   1,000

6. Next say, "Let's count by hundreds; but this time, let's start at forty." Begin a new vertical column on the board and write *40* at the top. Ask, "What is one hundred more than forty?" Continue, recording as students count.

Literature Connection

Count to a Million

The imaginative book *Count to a Million* by Jerry Pallotta inspires young counters to count higher than they ever thought possible. It begins, "If you can count to ten, you can count to one million."

Counting by Hundreds, Starting with 40

40

140

240

340

440

540

640

740

840

940

1,040

7. Compare the two columns of numbers (*Counting by Hundreds, Starting with 100* and *Counting by Hundreds, Starting with 40*). Ask students, "What do you notice when you compare the numbers in these two columns?"

Examples of Student Thinking

Kiko: "I saw that the forty stays the same going down, but the hundreds go one more each time."

Marcelina: "It's like when you count by tens. There's one more in the tens place each time then too."

8. Say, "Now let's count by two hundreds to one thousand." Record as students count:

   Counting by 200s to 1,000

   200

   400

   600

   800

   1,000

9. Ask students, "What do you notice in looking at the numbers in this column?" Discuss as appropriate.

Examples of Student Thinking

Malaya: "One thousand is ten hundreds. It goes up two, four, six, eight, ten."

Astrid: "So twelve hundreds is twelve hundred."

Pat: "And twenty hundreds is two thousand."

Teacher: "Turn and talk to your partner about what Pat means by twenty hundreds."

10. Draw students' attention to the number lines and introduce the game. If students have played *Race to 50* and *Race to 100*, remind them of those games and tell them that this game is called *Race to 1,000*.

11. Say, "This game is played in teams. There are two teams: a red team and a blue team." Divide the class in half and indicate that one side of the class will play for red and the other side will play for blue.

12. Continue, "The objective of the game is to get to or go over one thousand on your team's number line before the opposing team does the same on its number line. Teams take turns drawing action cards (point to the pile) to determine their move on the number line, beginning at zero."

13. If students have played *Race to 50* and *Race to 100*, ask them what they think will be the important numbers for the *Race to 1,000* action cards. Show them some of the action cards that include multiples of one hundred.

A Child's Mind . . .

Making Moves on the Number Line
Eventually students will keep track of their own jumps, but while modeling the game, it's best that you do the recording. Note that, at first, students may need to make jumps on the number line in increments smaller than one hundred. After several turns or games, the increments will likely increase. Encourage bigger jumps; also, encourage students to explain how they know they landed on and recorded the correct number.

14. Demonstrate a move by first choosing a student on the red team to draw an action card. Use questions to guide students in making the corresponding move on the number line. Let's say Mario draws the action card +200. Say, "Mario drew a plus-two hundred card, meaning he can move to two hundred. How could Mario get to two hundred? Who has another way? How do the red team members want to record their jump?" If the red team decides to take two jumps of one hundred for the plus-two hundred card, use the red marker to record the jumps on the corresponding number line:

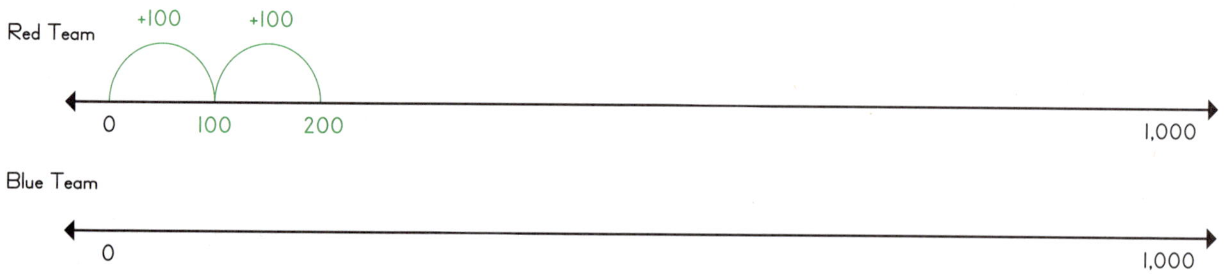

The red team's move after drawing the action card +200

15. Model a few rounds just using the action cards (teams take turns drawing a card, deciding their move, and recording their thinking on the corresponding number line). If a student draws a minus card that puts them backward before zero, they need to place it back in the card pile and draw again. After a card is drawn and played, put it back in the pile so it can be drawn randomly in the next turn.

16. Now introduce the use of the question cards. Explain that, after each team draws an action card and records the move on the number line, the team members must then draw a question card. Emphasize that a different team member should draw the question card, so as many team members as possible are participating in drawing the cards. Ask a student on the red team to draw a card and read the question to the team.

Teaching Tip

Spacing on the Open Number Line
It's important to pay attention to and demonstrate approximate spacing on an open number line. You obviously do not have an exact measurement of the distance between zero and one thousand, but students should have a sense of how the numbers increase along their location on the number line. You could say something like, "Well, your jump of two hundred would be about here on our number line. It's not quite halfway to five hundred, so we can mark it here. We're not worrying about being exact; we just want to show the jump of two hundred."

Example of Student Thinking

Note: In this example, the blue team is at 350 after drawing an action card with +300 and another with +50:

Milton: "We are at three hundred fifty. We need six hundred fifty more to get to one thousand. I know this because we need fifty more to get to four hundred and then six hundred to get to one thousand."

Differentiating Your Instruction

Supporting English Language Learners

The question cards may need to be omitted or modified in this game depending on students' ability to read. You may choose to have students work with just one question card and task each team with answering it each time they complete an action card. For example, you may decide to use the question: *How many more do I need to reach 1,000?* Students count forward on the number line to the target number and see that the difference can be obtained by counting on.

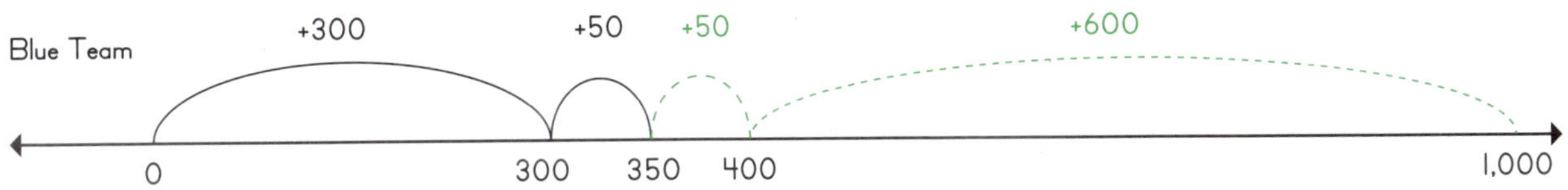

Blue team's jumps with Milton's comment.

17. Have students on both teams turn and talk about how to respond to the question. Then ask the red team to share its answer for their score so far.
18. Returning to our example, if the question card drawn had been *What is 100 more than what you have right now?*, a blue team member might respond, "One hundred more than three hundred fifty is four hundred fifty."
19. Now have both teams continue with a full turn using both a new action card and a new question card. Use key questions to support students in thinking about how the game works. For the jumps that go backward, record them below the number line, if necessary, to keep the recording from becoming too messy or confusing.

Math Matters!

Understanding Place Value

Really understanding what place value means is much more than knowing how to break a number into hundreds, tens, and ones. A child may be able to say that a number like 345 is three one hundreds, four tens, and five ones, but that child may not realize this number also has more than thirty tens or that this number is the same as 200 and 145. The child may not be comfortable thinking about how 345 is almost halfway between 300 and 400. It is also one hundred more than 245 and one hundred less than 445.

When children have a robust understanding of place value, they know whether a number is more than 10, 50, 100, 250, or another landmark number because of the value of the digits in each of the places.

20. Finally, introduce the use of the recording sheet (Reproducible 33) to the game. Support students in getting into the habit of writing down for each turn the number where they start, as well as the action card number. After students have recorded their thinking on their team's number line, task them with creating the corresponding equation. In our example, the blue team's recording sheet would look like this:

Starting Number	Acton Card Number	Result	Equation
0	+300	300	0 + 300 = 300
300	+50	350	300 + 50 = 350

21. Continue with a whole-class game, using the action cards, questions cards, and recording sheet, until a team makes it to 1,000 on the number line.

A Child's Mind . . .

Looking for Patterns

Some students will be ready to move comfortably to 1,000 on the number line whereas others are still becoming familiar with three-digit numbers. Find ways to help students look for patterns when counting. A chart counting by tens to 1,000 can be useful. It's often helpful to post a hundreds chart in the room for reference. (A reproducible hundreds chart is available in *It Makes Sense! Using Hundreds Charts to Build Number Sense* by Melissa Conkin and Stephanie Sheffield [Math Solutions, 2012], also in this series.)

Explore

22. When students are comfortable with the game procedures, have them play the game independently. Place them in teams of two and then pair up a red team with each blue team. Make sure each group has the number lines, red and blue markers, action cards, and question cards per the "Materials" list. As an alternative to piles, place each set of cards in a separate paper lunch sack and label each sack; teams draw from the sacks.

23. Encourage students to work together. Say, "When you play this game in teams of two, take turns with your teammate in choosing the action card and the question card, so you get a chance to do both tasks. Also, involve your partner by listening to their strategy before you record the move on your team's number line. After you've recorded a move, check each other's work. Does your recording match the action card?"

24. Help groups decide which team goes first. One way to do this is to have groups shake up their paper sacks of action cards and then have each team draw a card. The team that chooses the card with the greatest number on it goes first.

25. Circulate as students play, asking key questions to guide their thinking (see the Key Questions list on page 294).

Teaching Tip

Pairing Students

For this game, pair students with someone of like ability. Pairing students with similar mathematical skills may allow for more differentiation options and will afford optimum access to the mathematics involved in playing the game.

Math Matters!

Decomposing Numbers

What do students think about when they decompose numbers? Which numbers do they decompose easily? How do they know when to break apart a number? As teachers, we try to observe how children do these things so we can understand their levels of thinking. This is challenging work.

Sometimes students choose their favorite strategies that seem to make the best sense to them at the time. Some students are comfortable breaking numbers into parts that incorporate five as an addend: "Seven and eight is the same as five and five and two and three." Other students are comfortable with knowing their doubles and like to find ways to look for doubles or near doubles: "I know seven and seven are fourteen, so seven and eight is fifteen." Making a friendly ten is another way that students may try to add two numbers such as eight and seven: "Eight and two more are ten and then the other five will make fifteen."

Because we want students to have a wider range of successful strategies, we need to find ways for them to share their thinking with others. When students really listen to one another, they will try to follow the thinking of their peers and may even try some of the new strategies in their own problem situations. Instead of having only one way to solve a problem (counting on fingers or counting by ones, for example), we want students to have a wide variety of strategies and be flexible in thinking about a problem situation.

Summarize

26. After students have had the opportunity to play the game independently over a period of time (note that this game can be part of a math workshop learning station), plan a whole-class discussion.

27. Prepare for the discussion by showing the question card: *If you could make a new action card right now, what would it say?* Task students, in small groups, with thinking about the question in the context of the following sample game.

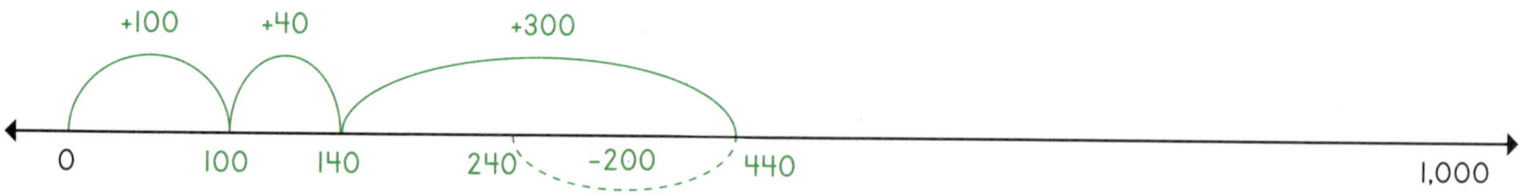

Race to 1,000 Recording Sheet

Starting Number	Action Card Number	Result	Equation
0	+100	100	0 + 100 = 100
100	+40	140	100 + 40 = 140
140	+300	440	140 + 300 = 440
440	-200	240	440 - 200 = 240

28. After students have completed the task in their small groups, bring them together as a whole class. Ask the question on the question card again: *If you could make a new action card right now, what would it say?* Use key questions to encourage students' thinking as well as to help students evaluate how their additional action card would impact the fun and fairness of the game.

Examples of Student Thinking

Misu: "I would put in a plus-seven hundred card so I could get closer to one thousand in one jump."

Kwanza: "There should be a plus-sixty card so we could get off the forty spot and get back on the even one hundred places."

Malaya: "We might as well go over one thousand with a plus-eight hundred card."

Teacher: "Talk with your group and see if you all can come up with any ideas about what might happen if we put these cards into the game. Would this be fair?" (The groups should talk among themselves, then reconvene as a whole class.)

Milton: "I don't think action cards with small jumps are very helpful."

Kaito: "I don't want one team to have such a big jump because then they'll win in only a few turns."

29. Make additional action cards using students' suggestions and include them with the original action cards sets. Have students play the game using these new sets. What are their findings? If we did this with the students' previous suggestions, they will likely discover that when a team draws the +700 or the +800 action card, the game is over all too soon. There likely won't be much of an impact with the addition of a +60 action card, however.

Teacher Reflection

When I asked second graders about how to improve the game *Race to 1,000* by designing new action cards, I was pleased to hear their suggestions. We actually tried some of their ideas to see if the game changed in any significant way. They decided that this game was no longer any fun when huge jumps (+800 or +700) were included in the action cards. They also thought that little jumps (such as adding or subtracting one-digit numbers) made the game too complicated. They never wanted to play with action cards that were not multiples of tens. I also asked them to read over the question cards and make suggestions for new cards. Some of the suggested additions were:

- What if you doubled the number on your action card, where would you land?
- Tell the other team what number you hope they get on their next action card.
- What would happen if you could only move to odd numbers?
- How would you play this game with three teams instead of two?
- Tell how you would play this game if you wanted to race to zero.

I appreciated working with these students. I believe that when they are given a chance to reflect on their experiences with these games, their learning is much more robust and meaningful.

Reproducibles

The following reproducibles are referenced and used with individual minilessons, lessons, and games throughout the text. These reproducibles are available in a downloadable, printable format. Visit www.mathsolutions.com/myonlineresources and register your product for access using the key code IMSNP. See page x for instructions.

(continued)

The following reproducible is referenced and used throughout the book:

The following Game Directions are referenced in individual games:

Number Cards

1	2	3
4	5	6
7	8	9

(continued)

(Number Cards, *continued*)

10	11	12
13	14	15
16	17	18
19	20	

Greater Than, Less Than, and Equal To Reference Chart

$>$

5 is greater than 3

$<$

4 is less than 7

 = 

5 is equal to 5

Ordinal Number Reference Chart

1	first	1st
2	second	2nd
3	third	3rd
4	fourth	4th
5	fifth	5th
6	sixth	6th
7	seventh	7th
8	eighth	8th
9	ninth	9th
10	tenth	10th

Dot Cards

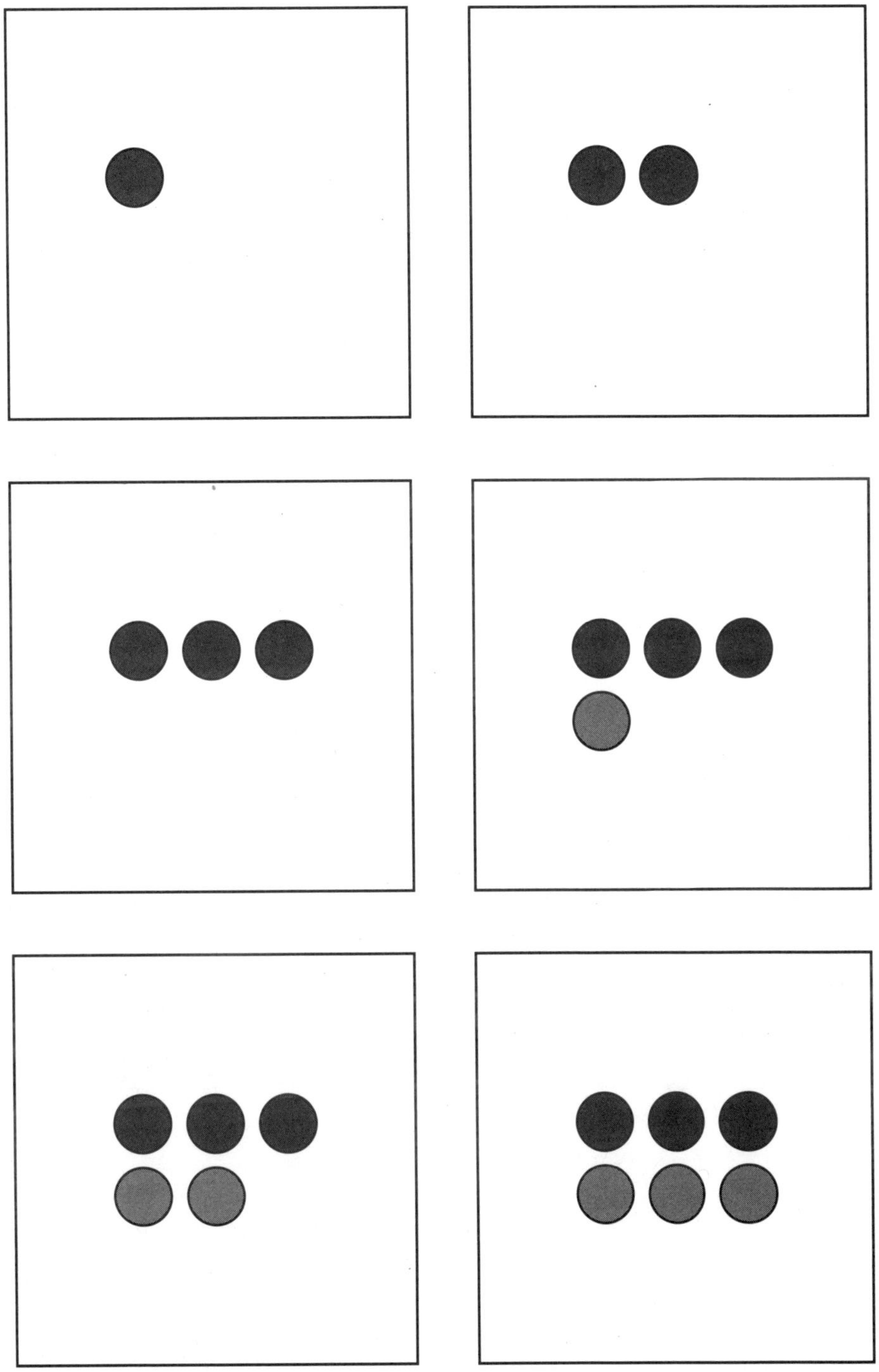

(continued)

(Dot cards 1–20, *continued*)

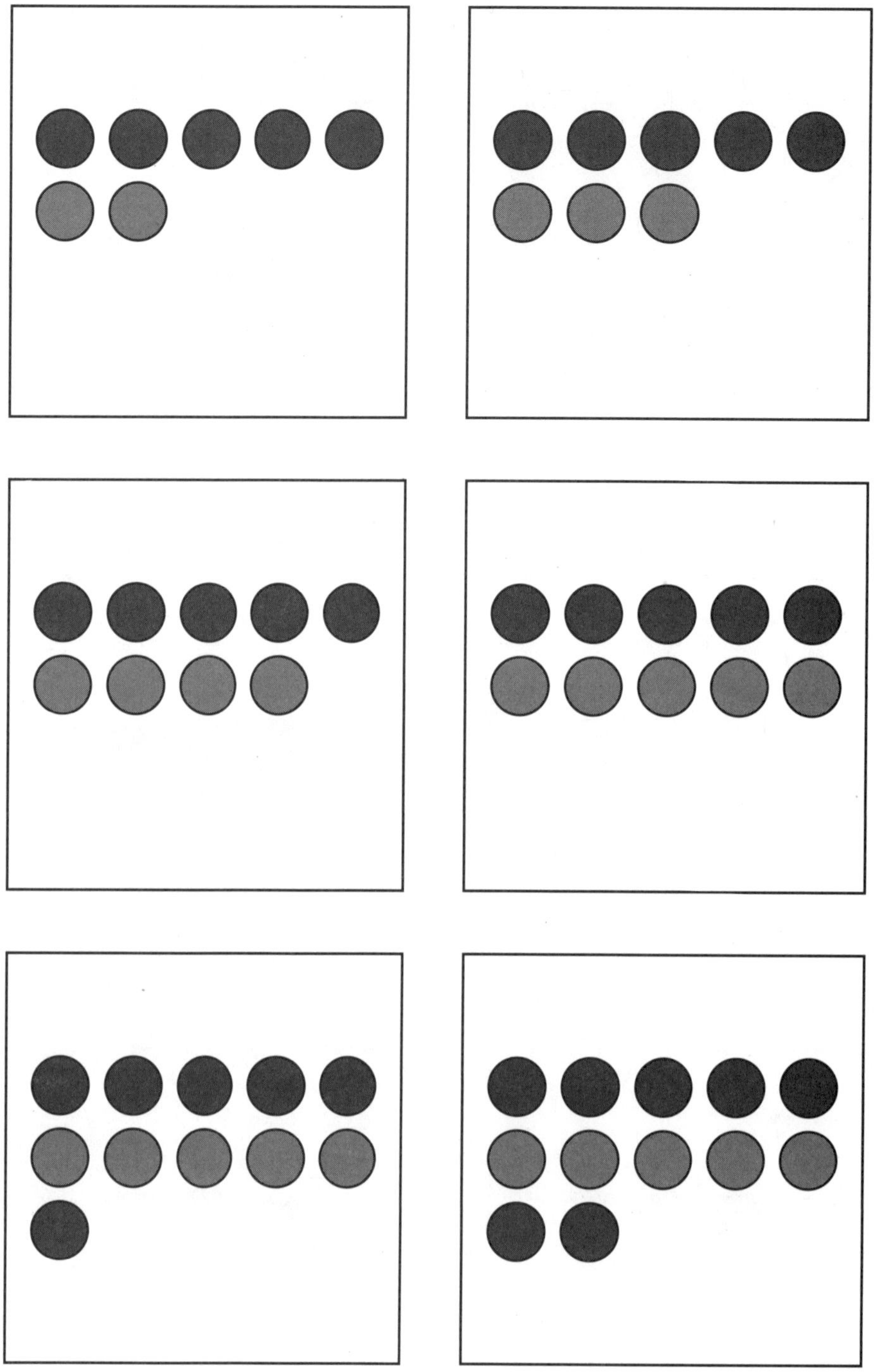

(continued)

(Dot cards 1–20, *continued*)

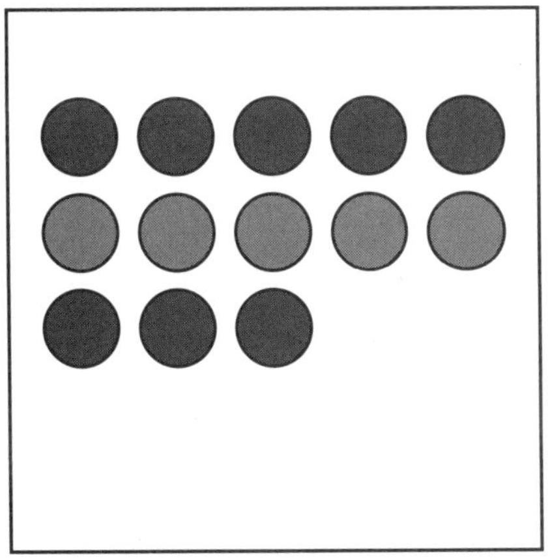

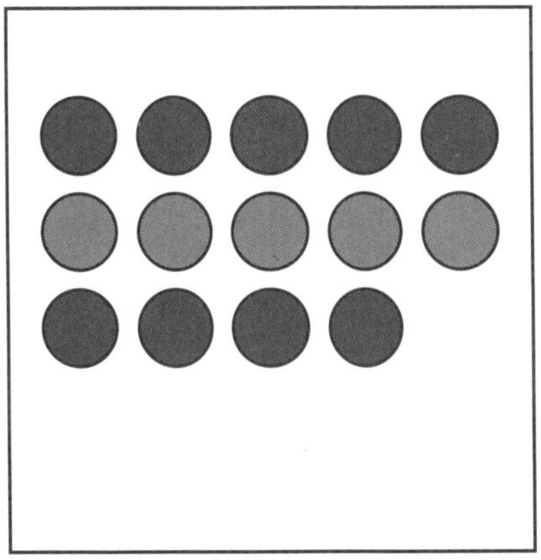

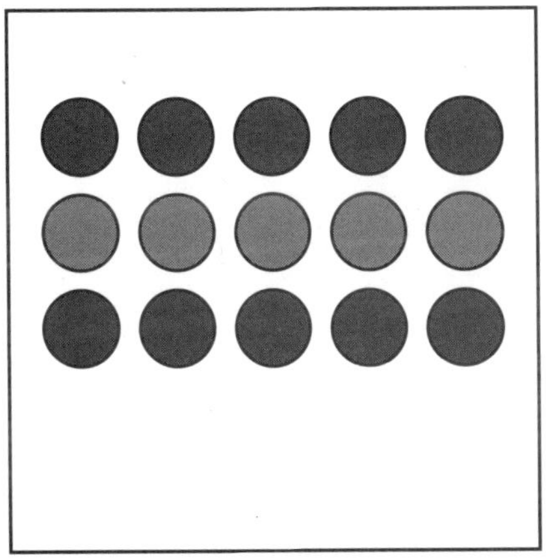

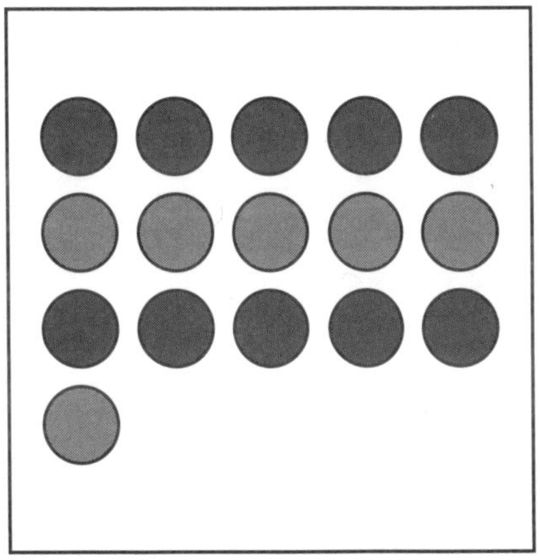

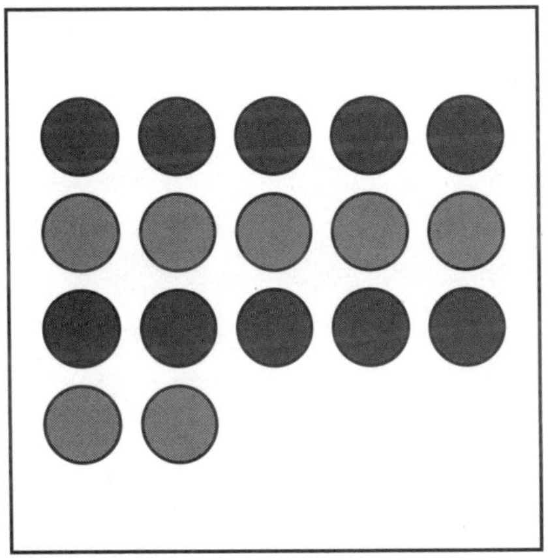

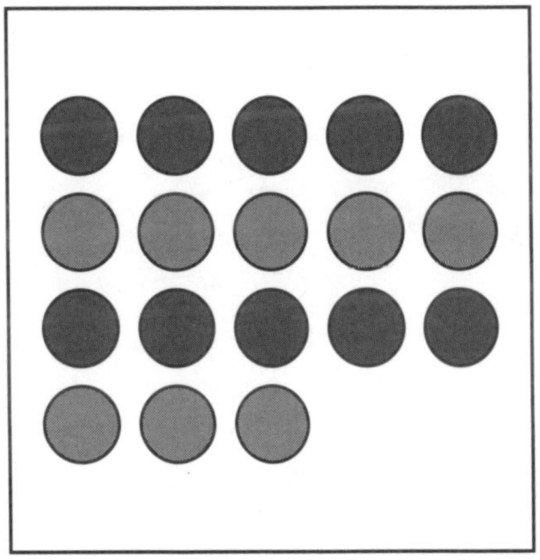

(continued)

(Dot cards 1–20, *continued*)

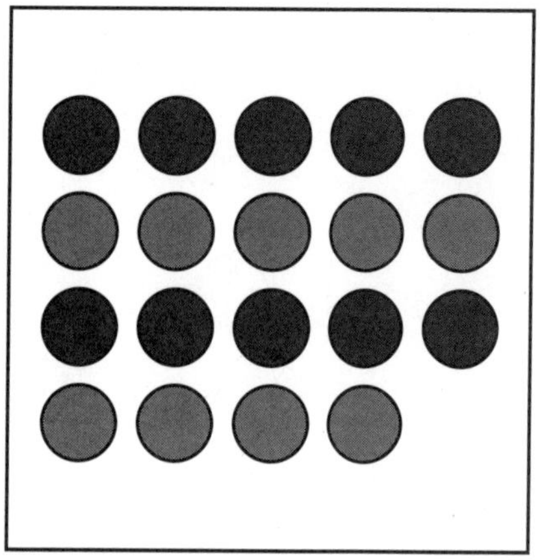

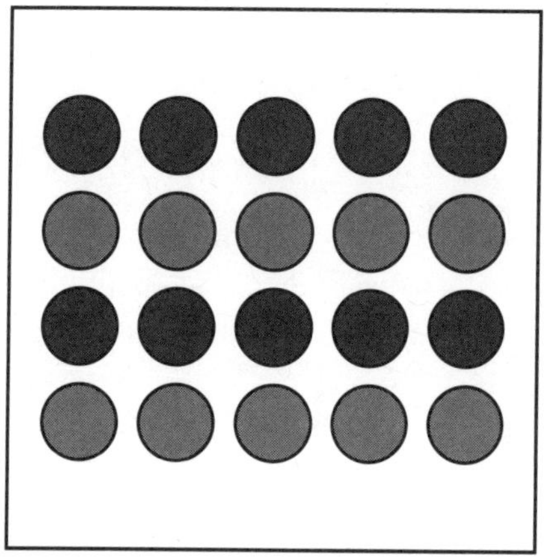

Jumping by Ones and Tens Number Cards

Cut the cards apart. One copy of this reproducible is one set of cards.

0	1	2
3	4	5
10	10	20

More/Less Spinner

Copy this spinner as needed for the lesson.

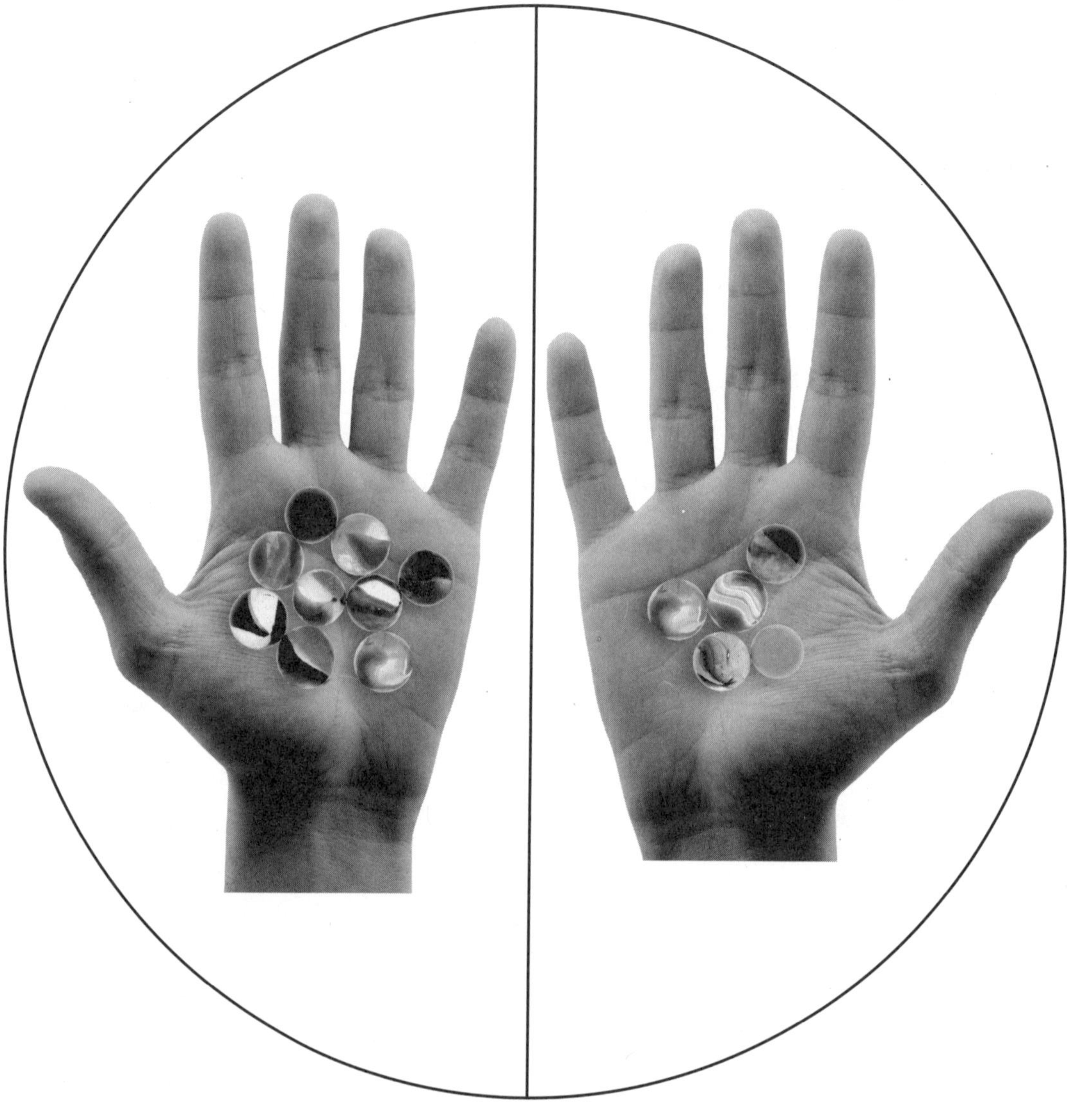

1. Pass out one large paper clip per pair of players.
2. Place the tip of a pencil in the middle of the spinner.
3. Slide the paper clip over the pencil so the clip lies flat on the spinner.
4. Spin the paper clip while holding the pencil or have a partner hold the pencil while you spin the paper clip.

Jumping by Ones, Fives, Tens, and Twenty Spinner

Copy this spinner as needed for the lesson.

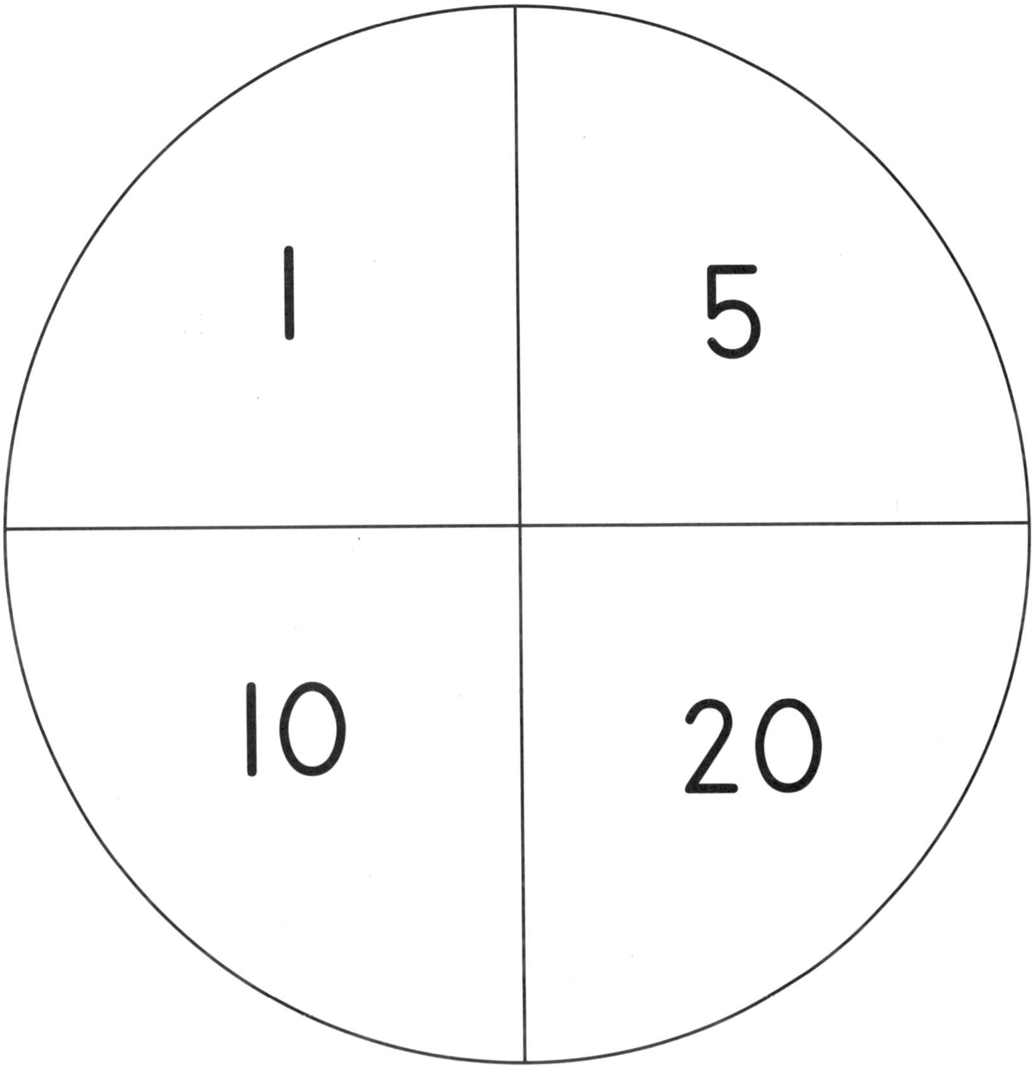

1. Pass out one large paper clip per pair of players.
2. Place the tip of a pencil in the middle of the spinner.
3. Slide the paper clip over the pencil so the clip lies flat on the spinner.
4. Spin the paper clip while holding the pencil or have a partner hold the pencil while you spin the paper clip.

Story Problems: Solving Story Problems Involving Missing Numbers at the End

Copy these story problems as needed. Cut each copy in half so there is one story problem per half page of paper.

Write your name on the blank line. Then, solve the problem using a number line or drawing.

Name: ________________

Three seagulls are sitting on a lunch table during recess. Nine more join them when ________________ (name of student) opens up a sandwich wrapper. How many seagulls are now sitting on the lunch table? What do you think happened?

Write your name on the blank lines. Then, solve the problem using a number line or drawing.

Name: ________________

The teacher has 50 pencils for the class. She sharpens 20 of them but has to stop because the bell is ringing. She asks ________________ (name of student) to please sharpen the rest of them while the students enter the classroom. How many pencils does ________________ (name of student) sharpen for the class? Are there be enough pencils for everyone?

Write your name on the blank line. Then, solve the problem using a number line or drawing.

Name: ____________________

Four bunnies are sitting on the grass next to ______________________________. Six more bunnies hop up to join them.
(name of student)

How many bunnies are now on the grass? What do the bunnies do now?

Write your name on the blank line. Then, solve the problem using a number line or drawing.

Name: ____________________

Nine apples are in a bowl on the table. ______________________________ eats two apples. How many apples are now in the
(name of student)

bowl? What can you make with the rest of the apples?

Story Problems: Solving Story Problems Involving Missing Numbers in the Middle

Copy these story problems as needed. Cut each copy in half so there is one story problem per half page of paper.

Name: ____________________

The teacher brings 4 pumpkins to school for the fall festival. Some children in the class bring in some more pumpkins. Now there are 10 pumpkins in the classroom. How many pumpkins did the children bring to school?

What can you do with these ten pumpkins for math time?

Name: ____________________

The teacher wants to have 4 students sit in the front row. There are 3 boys in the front row. How many girls should sit in the front row with these boys? Draw a picture of the front row and use B for boy and G for girl. How would you arrange the boys and girls? Why?

This is a picture frame for the front row.

(continued)

Name: ____________________

Four puppies are in a dog run. Some puppies are brought inside. Now there are 2 puppies in the dog run. How many puppies are inside?

What kind of puppies do you think they are?

Name: ____________________

Seven flowers are in a vase. Three flowers are red and the rest are white. How many flowers are white?

What kind of flowers do you think they are?

(continued)

Name: ___________________

Kevin wants to read 20 books for the library celebration in December. He already read 6 books during September. How many books will he need to read in October and November to meet his goal?

What favorite book would you recommend he read for the December event?

Name: ___________________

The teacher has forty pencils in a pencil jar. Fifteen are red and the rest are blue. How many blue pencils are there in the pencil jar. If she wants to have as many red pencils as blue pencils, what should she do?

Story Problems: Solving Story Problems Involving Missing Numbers at the Beginning

Copy these story problems as needed. Cut each copy in half so there is one story problem per half page of paper.

Name: ___________________

Some students in the classroom are making block buildings. Five more students join them on the floor to help. Now there are 11 students making block buildings. How many students were making block buildings on the floor before the other students joined them?

What kinds of buildings do you think they are making?

Name: ___________________

Some juice boxes are on a table in our classroom. The snack helper takes 17 juice boxes to the lunch area. Now there are 10 juice boxes on the table. How many juice boxes were on the table before?

Are there enough juice boxes for our whole class?

(continued)

Name: ___________________

The teacher wants to have 6 students sitting at each table. How many boys and how many girls can he put at one table? Make a list for the teacher, using B for boy and G for girl.

Why did you arrange the students the way you did?

Name: ___________________

Some bunnies are sitting on the grass. Six more bunnies hop up to join them. Now there are 10 bunnies sitting on the grass. How many bunnies were on the grass before the other bunnies joined them?

What do you think the bunnies will do?

(continued)

Name: ____________________

Some apples are in a bowl on the table. My brother ate 2 apples. Now there are 5 apples in the bowl. How many apples were in the bowl before my brother ate 2 of them?

Will there be enough apples for your entire family?

Name: ____________________

Grandma has 6 flowers. How many can she put in her red vase and how many can she put in her blue vase?

What kind of flowers do you think they are?

(continued)

Name: ___________________

We have ___ sticks of colored chalk for the chalk art festival on the blacktop outside our classroom. The principal brings more sticks of colored chalk for us to use. Now we have ___ sticks of chalk. How many sticks of chalk did the principal bring?

What should we draw on the blacktop for the chalk art festival?

More and Less Visual

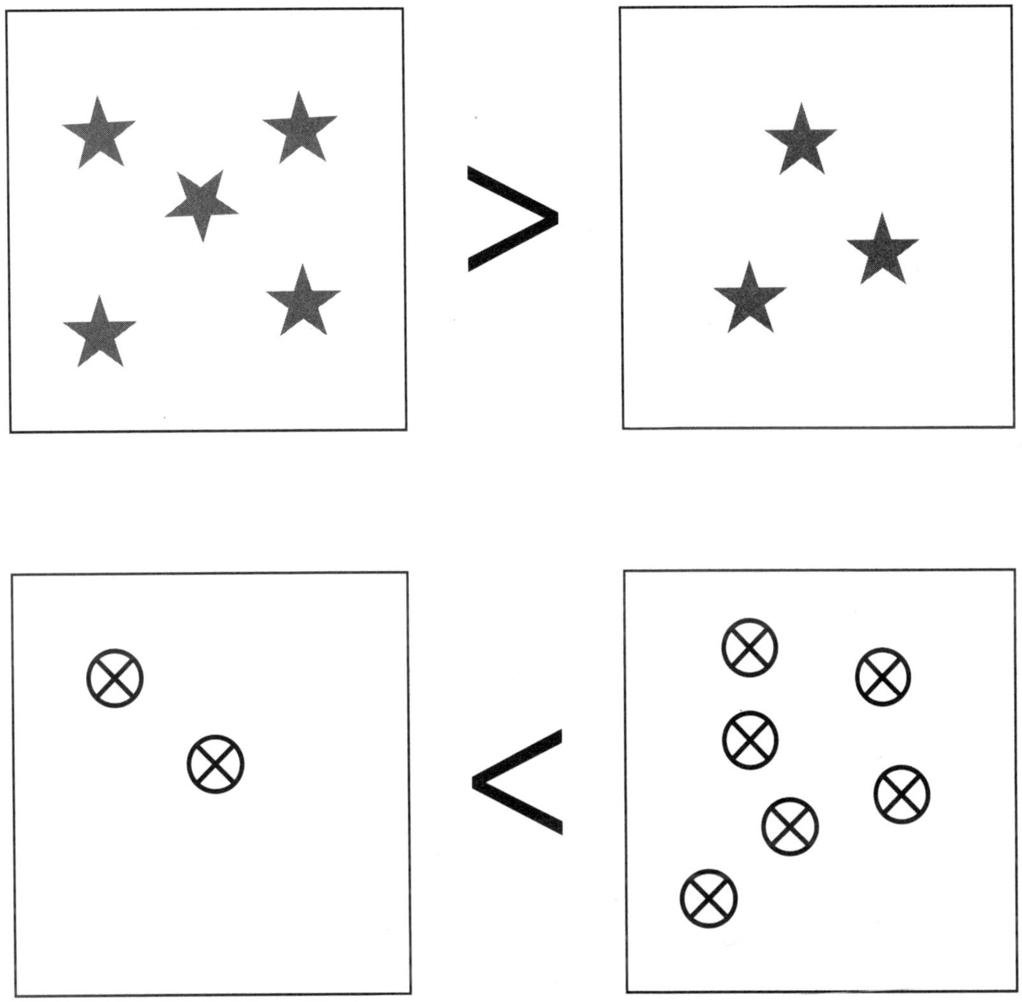

Story Problems: Solving Comparison Problems

Copy these story problems as needed. Cut each copy in half so there is one story problem per half page of paper.

Name: ________________

Write your name in the first blank space and a friend's name in the second blank space.

____________ has 4 dimes ______________ has 6 dimes. How many more dimes does ____________ have
(first person) (second person) (second person)

than ____________? If they pool their money, how much money do they have? What do you think they will buy?
(first person)

Name: ________________

The teacher has 9 new books from the book fair. The principal has 14 new books from the book fair. How many fewer books does the teacher have than the principal? What book would you want to buy from the book fair?

(continued)

Name: ________________

Write your name in the first blank space and a friend's name in the second blank space.

____________ (first person) has 6 more marbles than ____________ (second person). ____________ (second person) has 3 marbles.

How many marbles does ____________ (first person) have? How many do they have if they put their marbles together?

Name: ________________

Write your name in the first blank space and a friend's name in the second blank space.

____________ (first person) has 2 quarters. ____________ (second person) has 3 quarters. How many more quarters does ____________ (second person) have than ____________ (first person)? If they pooled their quarters, how much money would they have? What could they buy?

(continued)

Name: ____________________

The teacher has 4 new books from the book fair. The principal has 15 new books from the book fair. How many fewer books does the teacher have than the principal? Which new books should the teacher buy?

Name: ____________________

Write your name in the first blank space and a friend's name in the second blank space.

__________ (first person) has 6 more cards than ______________ (second person). ______________ (second person) has 9 cards. How many cards does __________ (first person) have? How many cards do they have if they put them together? Which card game should they play?

Race to 50 Action Cards

Cut the cards apart. One copy of this reproducible is one set of cards.

+10	+10	+10
+20	+5	+5
−10	+1	+1

Race to 50 Question Cards

Cut the cards apart. One copy of this reproducible is one set of cards.

How many more do you need to reach 50?	How far from 50 are you?	What is 10 *more* than what you have right now?
What is 10 *less* than what you have right now?	What is 5 *more* than what you have right now?	What is 5 *less* than what you have right now?
What is 1 *more* than what you have right now?	What is 1 *less* than what you have right now?	How far from 0 are you?

Race to 100 Action Cards

Cut the cards apart. One copy of this reproducible is one set of cards.

+10	+10	+10
+20	+5	+5
−10	−5	+1
+20	+40	+30

Race to 100 Question Cards

Cut the cards apart. One copy of this reproducible is one set of cards.

What is 10 *more* than what you have right now?	What is 10 *less* than what you have right now?	What is 5 *more* than what you have right now?
What is 5 *less* than what you have right now?	What is 1 *more* than what you have right now?	What is 1 *less* than what you have right now?
How far from 0 are you?	How many more do you need to reach 100?	How far from 100 are you?

(continued)

What is 20 *more* than what you have right now?	What is 20 *less* than what you have right now?	What is 15 *more* than what you have right now?
What is 15 *less* than what you have right now?	How far from 50 are you?	What is 10 *less* than what you have right now?

The Game of Pig on a Number Line Recording Sheet

Copy this recording sheet as needed to play the game, one sheet per game.

Game Directions

1. Roll two dice. Add the sum to your score on the number line.
2. If you roll one 1, you can't add the sum of your turn to your score.
3. If you roll two 1s, you start from 0 again.
4. You can take another turn if you don't have a 1.
5. Record your turn on the number line.
6. Pass the dice to the next player.
7. The player to reach 100 or more on a number line wins.

2	3	4	5	6	7	8	9	10	11	12

101 and Out! Recording Sheet

Game 1		tens	ones
Round 1	______		
Round 2 +	______		
Total	______		
Round 3 +	______		
Total	______		
Round 4 +	______		
Total	______		
Round 5 +	______		
Total	______		
Round 6 +	______		
Game Total	______		

Game 2		tens	ones
Round 1	______		
Round 2 +	______		
Total	______		
Round 3 +	______		
Total	______		
Round 4 +	______		
Total	______		
Round 5 +	______		
Total	______		
Round 6 +	______		
Game Total	______		

Game 3		tens	ones
Round 1	______		
Round 2 +	______		
Total	______		
Round 3 +	______		
Total	______		
Round 4 +	______		
Total	______		
Round 5 +	______		
Total	______		
Round 6 +	______		
Game Total	______		

Use the space below to record your numbers on the number line.

Hot Lava Bridge: Forward and Back, More and Less Action Cards

Cut the cards apart. One copy of this reproducible is one set of cards.

+10	+20	+30
+40	−10	−20
−30	+20	+1
−1	+10	+10
−10	−10	+1
+1	+1	−1
+0	−0	+20

Hot Lava Bridge: Forward and Back, More and Less Recording Sheet

Copy this recording sheet as needed to play the game, one sheet per game.

Example

Start	Action	End
0	+20	20

Start	Action	End

Game Directions

1. Choose an action card.
2. Move the figure to that space on the open number line.
3. Record your move on the recording sheet.
4. Next player repeats Steps 1–3.
5. The game ends when the figure moves off the hot lava bridge (i.e., number line) at either end, hence falling into hot lava!

Reproducible 21

The Larger Difference Game Recording Sheet

Copy this recording sheet as needed to play the game, one sheet per game.

Red Team, Student Names: ____________________ and ______________________

Blue Team, Student Names: ____________________ and ______________________

Round 1

The red team numbers are _________ and _________.

The difference between these numbers is _________.

The blue team numbers are _________ and _________.

The difference between these numbers is _________.

The team with the larger difference is: Red Team Blue Team
(*circle one and record the difference below*)

_____ > _____

(continued)

Round 2

The red team numbers are _________ and _________.

The difference between these numbers is _________.

The blue team numbers are _________ and _________.

The difference between these numbers is _________.

The team with the larger difference is: Red Team Blue Team
(*circle one and record the difference below*)

_____ > _____

(continued)

Round 3

The red team numbers are _________ and _________.

The difference between these numbers is _________.

The blue team numbers are _________ and _________.

The difference between these numbers is _________.

The team with the larger difference is: Red Team Blue Team (*circle one*)

_____ > _____

The Smaller Difference Game Recording Sheet

Copy this recording sheet as needed to play the game, one sheet per game.

Red Team, Student Names: ____________________ and ____________________

Blue Team, Student Names: ____________________ and ____________________

Round 1

The red team numbers are ________ and ________.

The difference between these numbers is ________.

The blue team numbers are ________ and ________.

The difference between these numbers is ________.

The team with the smaller difference is: Red Team Blue Team
(*circle one and record the difference below*)

_____ < _____

(continued)

Round 2

The red team numbers are _________ and _________.

The difference between these numbers is _________.

The blue team numbers are _________ and _________.

The difference between these numbers is _________.

The team with the smaller difference is: Red Team Blue Team
(*circle one and record the difference below*)

_____ < _____

(*continued*)

Round 3

The red team numbers are ________ and ________.

The difference between these numbers is ________.

The blue team numbers are ________ and ________.

The difference between these numbers is ________.

The team with the smaller difference is: Red Team Blue Team
(*circle one and record the difference below*)

_____ < _____

(continued)

Round 4

The red team numbers are ________ and ________.

The difference between these numbers is ________.

The blue team numbers are ________ and ________.

The difference between these numbers is ________.

The team with the smaller difference is: Red Team Blue Team
(*circle one and record the difference below*)

_____ < _____

(continued)

Round 5

The red team numbers are __________ and __________.

The difference between these numbers is __________.

The blue team numbers are __________ and __________.

The difference between these numbers is __________.

The team with the smaller difference is: Red Team Blue Team
(*circle one and record the difference below*)

_____ < _____

Adding Nines, Tens, and Elevens Spinner

Copy this spinner as needed for the game.

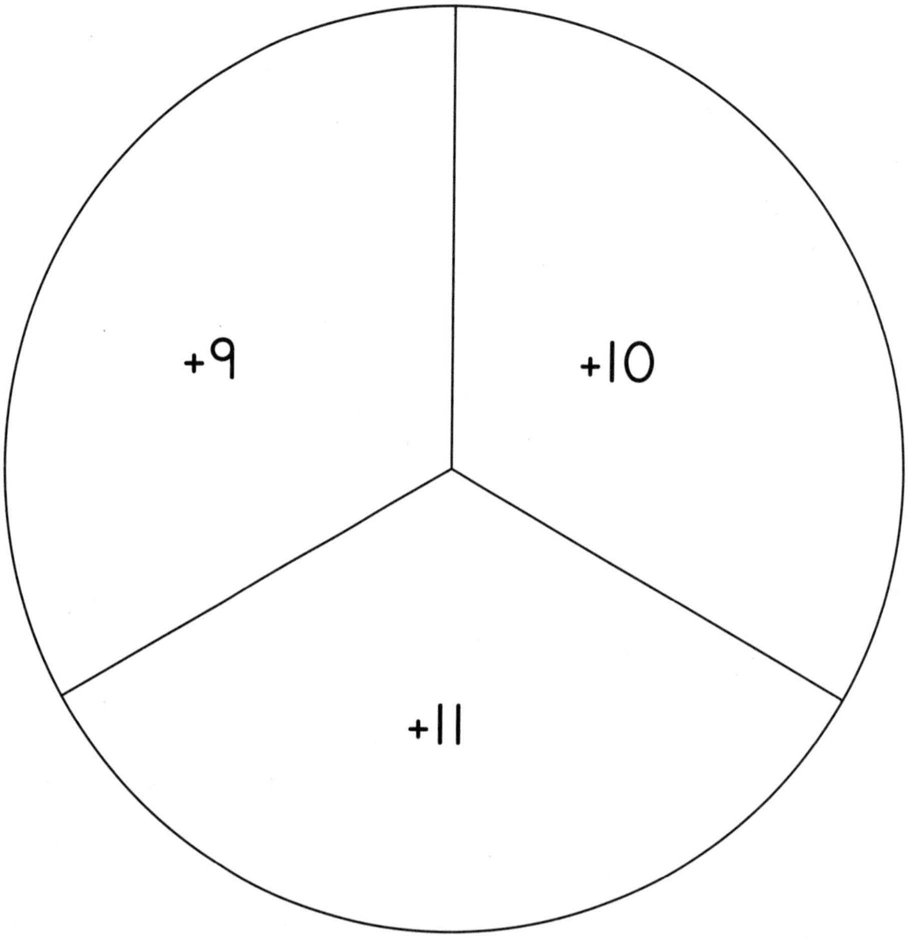

1. Pass out one large paper clip and one pencil per pair of players.
2. Use the tip of the pencil to keep the paper clip on the spinner.
3. Spin the paper clip while holding the pencil or have a partner hold the pencil while you spin the paper clip.

Get to the Target (Version 1) Spinner

Copy this spinner as needed for the game.

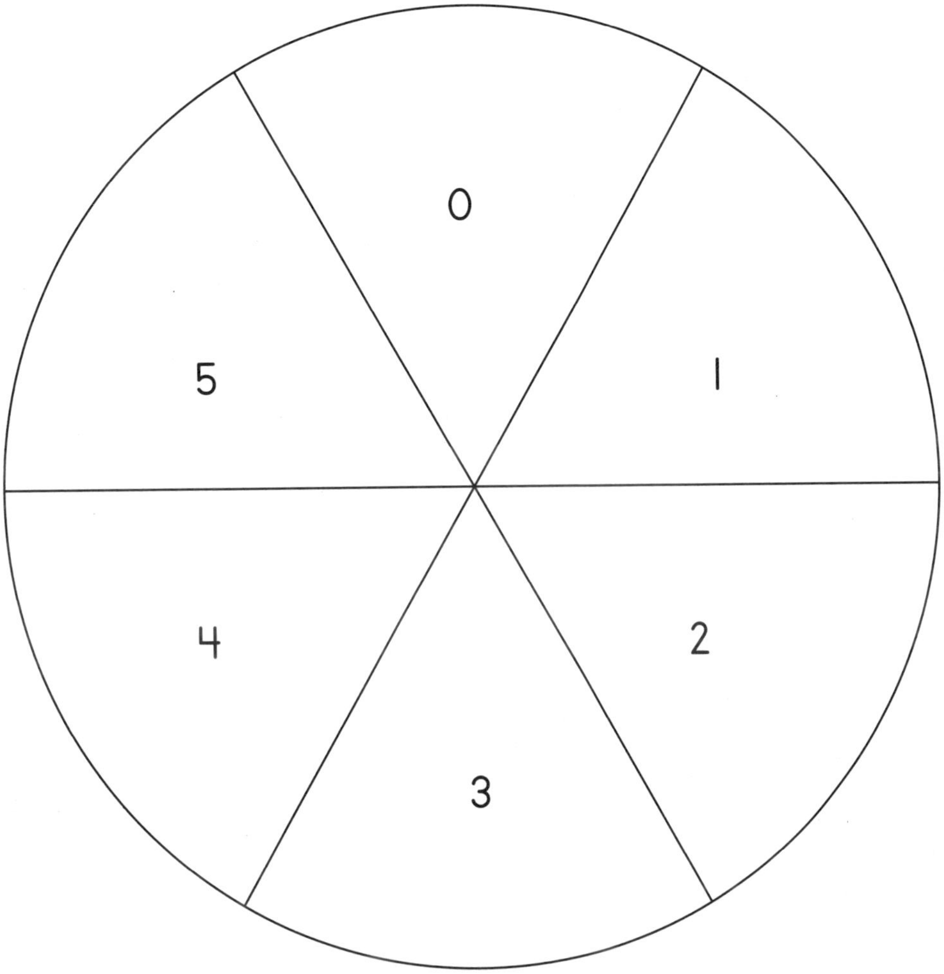

1. Pass out one large paper clip and one pencil per pair of players.
2. Use the tip of the pencil to keep the paper clip on the spinner.
3. Spin the paper clip while holding the pencil or have a partner hold the pencil while you spin the paper clip.

Get to the Target (Version 1) Recording Sheet

Copy this recording sheet as needed to play the game, one sheet per game.

PARTNER A: ______________________________

Our Target Number: ________

Starting at Zero or Starting at 60 (*circle one*)

My Number Line and Thinking

Total Number of Jumps: ________

Less Jumps Than My Partner or **More Jumps Than My Partner** (*circle one*)

PARTNER B: ______________________________

Our Target Number: ________

Starting at Zero or **Starting at 60** (*circle one*)

My Number Line and Thinking

Total Number of Jumps: ________

Less Jumps Than My Partner or More Jumps Than My Partner (*circle one*)

Sentence Frames Chart

Enlarge and display this chart to support students in articulating their thinking.

Sentence Frames

First, I started at ________.

Then I jumped ________ to ________.

Next, I jumped ________ more (or less) to ________.

Then I jumped ________ more (or less) to ________.

Finally, I jumped ________ to the target number of ________.

Get to the Target (Version 2) Spinner

Copy this spinner as needed for the game.

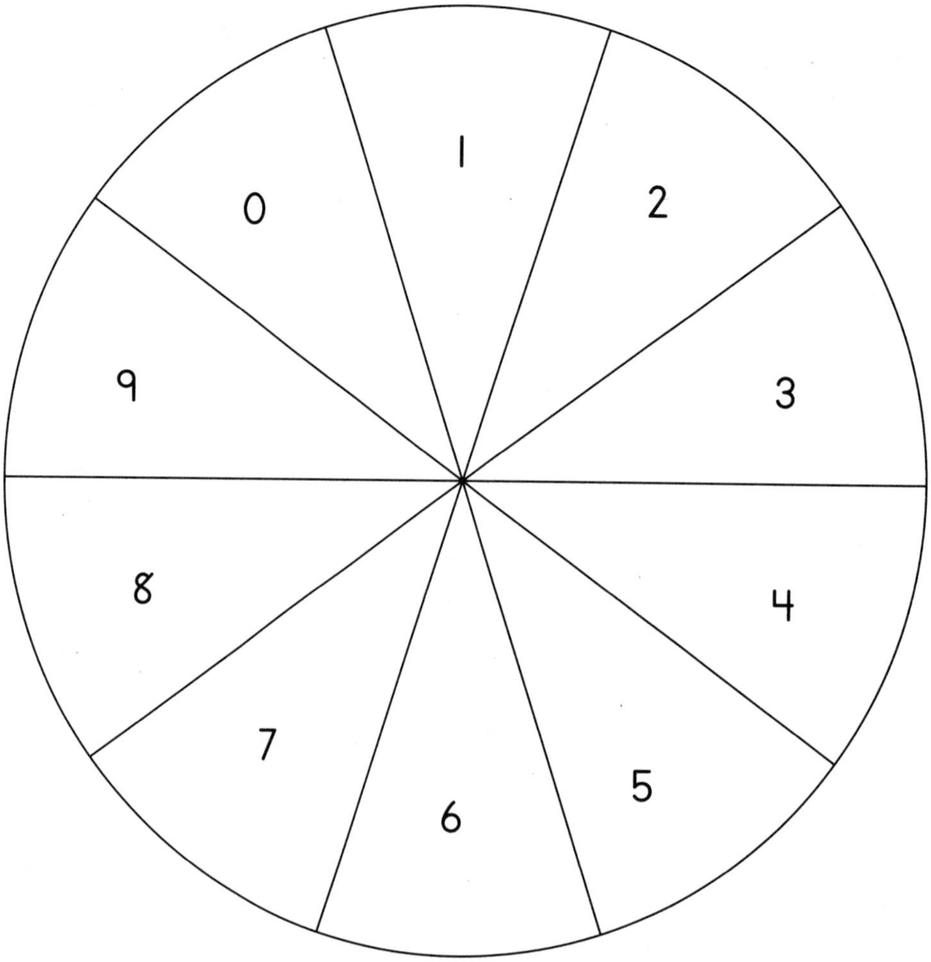

1. Pass out one large paper clip and one pencil per pair of players.
2. Use the tip of the pencil to keep the paper clip on the spinner.
3. Spin the paper clip while holding the pencil or have a partner hold the pencil while you spin the paper clip.

Get to the Target (Version 2) Recording Sheet

Copy this recording sheet as needed to play the game, one sheet per game.

PARTNER A: ______________________________

Our Target Number: ________

Starting at Zero or **Starting at 100** (*circle one*)

My Number Line and Thinking

Total Number of Jumps: ________

Less Jumps Than My Partner or **More Jumps Than My Partner** (*circle one*)

PARTNER B: ______________________________

Our Target Number: ________

Starting at Zero or **Starting at 100** (*circle one*)

My Number Line and Thinking

Total Number of Jumps: ________

Less Jumps Than My Partner or **More Jumps Than My Partner** (*circle one*)

Get to the Target (Version 3) Spinner

Copy this spinner as needed for the game.

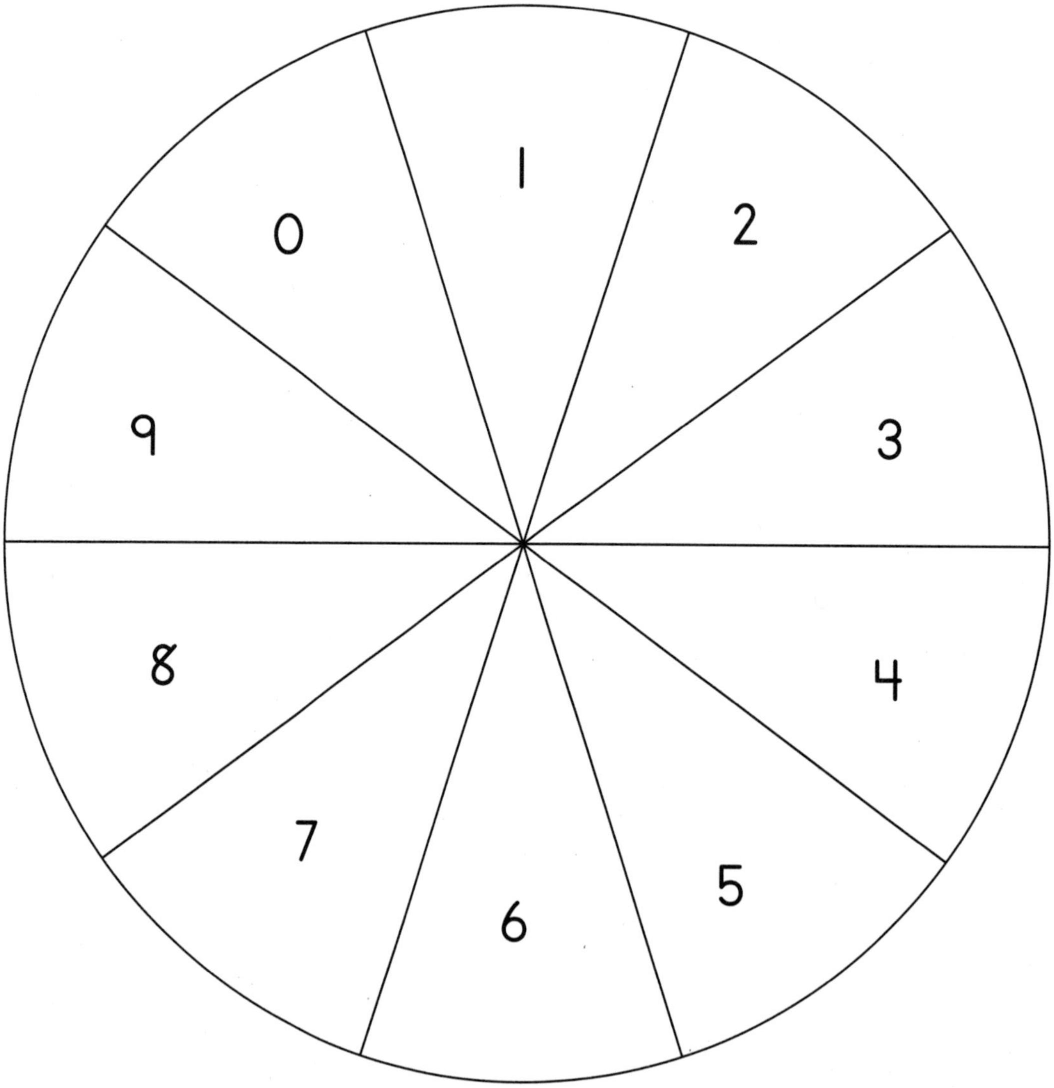

1. Pass out one large paper clip and one pencil per pair of players.
2. Use the tip of the pencil to keep the paper clip on the spinner.
3. Spin the paper clip while holding the pencil or have a partner hold the pencil while you spin the paper clip.

Get to the Target (Version 3) Recording Sheet

Copy this recording sheet as needed to play the game, one sheet per game.

PARTNER A: ______________________________

Our Target Number: ________

Starting at Zero or **Starting at 1,000** (*circle one*)

My Number Line and Thinking

Total Number of Jumps: ________

Less Jumps Than My Partner or **More Jumps Than My Partner** (*circle one*)

PARTNER B: ______________________________

Our Target Number: ________

Starting at Zero or **Starting at 1,000** (*circle one*)

My Number Line and Thinking

Total Number of Jumps: ________

Less Jumps Than My Partner or **More Jumps than My Partner** (*circle one*)

Race to 1,000 Action Cards

Cut the cards apart. One copy of this reproducible is one set of cards.

+100	+200	+300
+100	+100	+200
+50	+40	-100

(continued)

+50	+50	+30
-200	-50	+10
+20	+30	-10

Race to 1,000 Question Cards

Cut the cards apart. One copy of this reproducible is one set of cards.

How many more do you need to reach 1,000?	How far from 500 are you?	What is 100 *more* than what you have right now?
What is 100 *less* than what you have right now?	What is 50 *more* than what you have right now?	What is 50 *less* than what you have right now?
What is 5 *more* than what you have right now?	What is 5 *less* than what you have right now?	What is the largest jump you have made so far?
What is the smallest jump you have made so far?	How far ahead or behind are you in this game?	If you could make a new action card right now, what would it say?

Race to 1,000 Recording Sheet

Copy this recording sheet as needed to play the game, one sheet per game.

Example

Starting Number	Action Card Number	Result	Equation
100	+200	300	100 + 200 = 300
300	+40	340	300 + 40 = 340

Starting Number	Action Card Number	Result	Equation

How to Make an Open Number Path

Materials:

paper cash register tape

10 red and 10 yellow paper squares (3 inches by 3 inches)

glue stick or stapler

Directions:

1. Roll out a length of paper cash register tape that is approximately 60 inches long.
2. Affix 5 red paper squares with glue or stapler, next to one another in a row.
3. Now affix 5 yellow paper squares next to the row of red squares.
4. Affix 5 more red paper squares next.
5. Finish with 5 more yellow paper squares.

Other lengths can be made with 30 red and 30 yellow paper squares and so on.

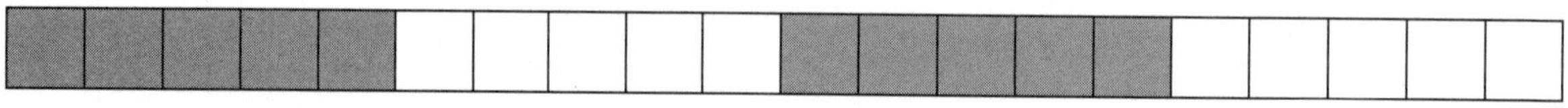

Open number path.

Game Directions

Mystery Number (Version 1)

Objective

Players try to identify a mystery number on a number path using a series of guesses. After each guess the players make, the holder of the mystery number gives a clue using the words *less* or *more* and *less than* or *greater than*. They then use a sticky note to cover the guessed number. The game is over when the correct number is guessed.

Materials

number path labeled *1* through *10* or *1* through *20*, with the numbers spaced equally apart (approximately 3 inches), created using paper sentence strips or cash register paper that is approximately 3 inches tall

sticky notes (approximately 3 inches by 3 inches)

Greater Than, Less Than, and Equal To Reference Chart (Reproducible 2)

Players

2 or more

Directions

1. Determine which player or team gets to choose the mystery number—this is Player 1.
2. Player 1 chooses a number on the number path. This is the mystery number; do not share it. Write the mystery number on a piece of paper and turn the paper over so the other players can't see it.
3. The other player(s) take turns guessing the number.
4. After each turn, Player 1 gives clues with the words *less* or *more* and *less than* or *greater than*.
5. Player 1 covers each guessed number with a sticky note.
6. When the mystery number is guessed, Player 1 congratulates the guessers.
7. Players switch roles and play again.

Variation

Players use an open number path, created with red and yellow paper squares (approximately 3 inches by 3 inches), ten of each color, organized in groups of five starting with red. Players use clothespins or sticky notes with an arrow drawn facing left on one and facing right on the other and move the clothespins according to the guesses.

Game Directions

Mystery Number (Version 2)

Objective

Players try to identify a mystery number on an open number path using a series of guesses. After each guess the players make, the holder of the mystery number gives a clue using the words *less* or *more* and *less than* or *greater than.* They then use two sticky notes with an arrow drawn facing left on one and facing right on the other to narrow the range of possible numbers. The game is over when the correct number is guessed.

Materials

open number path, created using cash register tape or sentence strips (enough for an open number path to 40)

red and yellow paper squares (approximately 3 inches by 3 inches), 20 of each color, organized in groups of five starting with red

2 sticky notes with an arrow drawn facing left on one and facing right on the other

Greater Than, Less Than, and Equal to Reference Chart (Reproducible 2)

Players

2 or more

Directions

1. Determine which player or team gets to choose the mystery number—this is Player 1.
2. Player 1 chooses a number on the open number path. This is the mystery number; do not share it. Write the mystery number on a piece of paper and turn the paper over so the other players can't see it.
3. The other player(s) take turns guessing the number.
4. After each turn, Player 1 gives clues with the words *less* or *more* and *less than* or *greater than.*
5. Player 1 places sticky notes with arrows facing left and right above each guessed number to show that the mystery number is less than or greater than that number.
6. When the mystery number is guessed, Player 1 congratulates the guessers.
7. Players switch roles and play again.

Game Directions

Mystery Number (Version 3)

Objective

Players try to identify a mystery number on an open number line using a series of guesses. After each guess the players make, the holder of the mystery number gives a clue using the words *less* or *more* and *less than* or *greater than*. They then use two sticky notes with an arrow drawn facing left on one and facing right on the other to narrow the range of possible numbers. The game is over when the correct number is guessed.

Materials

open number line labeled *0* and *100*, created on a blank piece of paper

2 sticky notes with an arrow drawn facing left on one and facing right on the other

Greater Than, Less Than, and Equal To Reference Chart (Reproducible 2)

Players

2 or more

Directions

1. Determine which player or team gets to choose the mystery number—this is Player 1.
2. Player 1 chooses a number on the number line. This is the mystery number; do not share it. Write the mystery number on a piece of paper and turn the paper over so the other players can't see it.
3. The other player(s) take turns guessing the number.
4. After each turn, Player 1 records the guess on the number line and gives clues with the words *less* or *more* and *less than* or *greater than*.
5. Player 1 places sticky notes with arrows facing left and right above each guessed number to show that the mystery number is less than or greater than that number.
6. When the mystery number is guessed, Player 1 congratulates the guessers.
7. Players switch roles and play again.

Game Directions

Race to 50

Objective

Teams take turns drawing a *Race to 50* action card. The direction on the action card tells the players how far to move on the number line (adding or subtracting one, five, ten, or twenty). After each move, the team then draws a *Race to 50* question card and answers the question on the card. The first team to get to fifty or beyond wins.

Materials

open number lines labeled *0* and *50*, 1 marked *Red Team* and 1 marked *Blue Team*, created on a blank piece of paper

2 colored pencils or markers (1 red and 1 blue)

Race to 50 Action Cards (Reproducible 13), 1 set (made from 1 copy of the reproducible)

Race to 50 Question Cards (Reproducible 14), 1 set (made from 1 copy of the reproducible)

2 paper lunch sacks, 1 labeled *Action Cards* and 1 labeled *Question Cards* (to hold the cards)

Players

2 or more

Directions

1. Determine who is the red team and who is the blue team. Each team draws an action card to determine which team goes first (the team with the larger number starts; if a team gets the –10 card they choose again). For the purposes of these directions, the red team will go first.
2. A player from the red team draws an action card, records the move on their number line, and returns the card to the paper lunch sack.
3. A player on the red team draws a question card, answers the question, and returns the card to the paper lunch sack. If there is more than one person on each team, players take turns choosing the action card and the question card. The red team passes the paper lunch sacks to the blue team.
4. A player from the blue team draws an action card, records the move on their number line, and returns the card to the paper lunch sack.
5. A player on the blue team draws a question card, answers the question, and returns the card to the paper lunch sack. If there is more than one person on each team, players take turns choosing the action card and the question card. The blue team passes the paper lunch sacks back to the red team.
6. The red team repeats the steps, adding more jumps onto their number line. The blue team also repeats this process.
7. The first team to make jumps to or beyond 50 on the number line wins.

Variations

Extend the number line to 100 and play *Race to 100*. Or play *Race from 100 to 200* using a 100-to-200 number line. For both *Race to 100* and *Race from 100 to 200*, use Reproducibles 15 and 16.

Game Directions

The Game of Pig on a Number Line

Objective

Players alternate turns, rolling two dice to determine how far to move on their individual number lines. They may roll the dice as many times as they want during their turn, building on their last move each time. However, there may be penalties per the game constraints if players get too greedy (in other words, become a pig)! The first player to reach 100 on their number line wins.

Game Constraints

If you roll one 1, you must take away any moves you've made during that turn.

If you roll two 1s, you must remove all your moves and start from 0 again.

Materials

open number line labeled *0* and *100*, 1 per player

2 dice (labeled *1–6*)

2 or 3 markers (in different colors)

Players

2 or more

Directions

1. Determine who will be Player 1 (or Team 1) and Player 2 (or Team 2).
2. Player 1 rolls the dice, adds the rolled numbers together, and records, on their number line, a jump equal to the sum. If multiple players are on each team, each player should record their jump in a different color. Player 1 then decides whether to roll again and again, until they decide to stop or until they come across one of the game constraints. When Player 1's turn is over, they pass the dice to Player 2.
3. Player 2 rolls the dice, adds the rolled numbers together, and records, on their number line, a jump equal to the sum. If multiple players are on each team, each player should record their jump in a different color. Player 2 then decides whether to roll again and again, until they decide to stop or until they come across one of the game constraints. When Player 2's turn is over, they pass the dice back to Player 1.
4. Player 1 repeats the steps, adding more jumps onto their number line. Player 2 also repeats this process.
5. Play continues until one player reaches or goes beyond 100 on their number line. This player is the winner.

Game Directions

101 and Out!

Objective

Players roll a die and decide whether the number should be placed in the ones or tens place. For example, if a five is rolled, it could be designated as five (ones place) or fifty (tens place). After six rolls, the player that is closer to one hundred without going over is the winner.

Materials

open number line labeled *0* and *100*

1 dice (labeled *1–6*)

101 and Out! Recording Sheet (Reproducible 18), 1 per player

pencils, 1 per player

Players

2

Directions

1. Determine who will be Player 1 and Player 2.
2. Player 1 rolls the die, decides whether they will put the number in the tens place or the ones place, and records the move on their number line. After recording the number on the recording sheet, Player 1 passes the die to Player 2.
3. Player 2 rolls the die, decides whether they will put the number in the tens place or the ones place, and record the move on the number line. After recording the number on the recording sheet, Player 2 passes the die back to Player 1.
4. Player 1 repeats the steps, this time adding the newly created number to the number from Round 1 and recording the total. Player 2 also repeats this process.
5. Play continues until both players have rolled the die six times. Players must use each number rolled. If a roll causes a player to pass 100 on the number line, they are disqualified (hence the game's title: 101 and Out!).
6. Players 1 and 2 compare their game total to see who is closer to one hundred without going over. This person is the winner.

Variation

This game can also be played with two groups of two players.

Game Directions

Hot Lava Bridge: Forward and Back, More and Less

Objective

Teams take turns drawing a *Hot Lava Bridge: Forward and Back, More and Less* action card. The direction on the action card tells the players how far to move the action figure along the hot lava bridge (i.e., number line). The game ends when one of the players moves the action figure off the number line at either end (hence, falling into hot lava).

Extension

Players draw two cards at a time and combine the numbers into one move on the number line.

Materials

large open number line labeled *0* and *50* (on approximately 18-inch-by-24-inch paper)

action figure, chess piece, or other small toy

2 markers (in different colors)

Hot Lava Bridge: Forward and Back, More and Less Action Cards (Reproducible 19), 1 set (made from 1 copy of the reproducible)

1 paper lunch sack (to hold the cards)

Hot Lava Bridge: Forward and Back, More and Less Recording Sheet (Reproducible 20)

Players

4

Directions

1. Determine who is on Team 1 and who is on Team 2. Each team draws an action card to determine which team goes first (the team with the larger number starts; if a team draws a minus card they choose again). For the purpose of these directions, Team 1 will go first.
2. Team 1 draws an action card, moves the action figure to that space on the open number line, and records the move on the number line and recording sheet. Team 1 passes the paper lunch sack to Team 2.
3. Team 2 draws an action card, moves the action figure to that space on the open number line, and records the move on the number line and recording sheet. Team 2 passes the paper lunch sack back to Team 1.
4. Team 1 repeats the steps, moving the action figure along the open number line. Team 2 also repeats this process. Team members take turns choosing the action card and moving the figure.
5. The game ends when a team moves the figure off the hot lava bridge (i.e., number line) at either end, hence falling into hot lava!

Game Directions

The Larger Difference Game

Objective

Opposing teams roll two dice and create two two-digit numbers. The teams find the difference between the numbers using jumps on a number line and compare to determine who found the larger difference.

Materials

2 colored pencils or markers (1 red and 1 blue)

2 dice (labeled *1* through *6*)

The Larger Difference Game Recording Sheet (Reproducible 21)

Players

4

Directions

1. Determine who will be on the red team and who will be on the blue team. Roll the dice to determine which team goes first (the team with the larger roll starts). For the purposes of these directions, the red team will go first.
2. A player from the red team rolls the dice, creates two, two-digit numbers and records those numbers on the recording sheet. (If the same two numbers are rolled, the player needs to roll again.)
3. The other player on the red team finds the difference between the two numbers using jumps on their teams' number line. The red team passes the dice to the blue team.
4. A player from the blue team rolls the dice, creates two two-digit numbers and records those numbers on the recording sheet. (If the same two numbers are rolled, the player needs to roll again.)
5. The other player on the blue team finds the difference between the two numbers using jumps on their teams' number line.
6. Both teams use the same recording sheet. The winner of each round is the team with the largest difference.

Variations

Drawing Two Number Cards 1 through 9

Players use number cards labeled *1* through *9* (Reproducible 1) instead of two dice. Players draw two cards from a paper lunch sack, create two two-digit numbers from the cards drawn, and then return the cards to the sack.

Using Three Dice Instead of Two

Players use three dice instead of two dice. Players roll all three dice and create two three-digit numbers. If two of the three dice rolled have the same number, players must roll the three dice again until all three are different numbers.

Drawing Three Number Cards 1 through 9

Players use number cards labeled *1* through *9* (Reproducible 1) instead of two dice. Players draw three cards from a paper lunch sack, create two, three-digit numbers from the cards drawn, and then return the cards to the sack.

Game Directions

The Smaller Difference Game

Objective

Opposing teams roll two dice and create two, two-digit numbers. The teams find the difference between the numbers using jumps on a number line and compare to determine who found the smaller difference.

Materials

2 colored pencils or markers (1 red and 1 blue)

2 dice (labeled *1* through *6*)

The Smaller Difference Game Recording Sheet (Reproducible 21)

Players

4

Directions

1. Determine who will be on the red team and who will be on the blue team. Roll the dice to determine which team goes first (the team with the smaller roll starts). For the purposes of these directions, the red team will go first.
2. A player from the red team rolls the dice, creates two, two-digit numbers and records those numbers on the recording sheet. (If the same two numbers are rolled, the player needs to roll again.)
3. The other player on the red team finds the difference between the two numbers using jumps on their team's number line. The red team passes the dice to the blue team.
4. A player from the blue team rolls the dice, creates two, two-digit numbers and records those numbers on the recording sheet. (If the same two numbers are rolled, the player needs to roll again.)
5. The other player on the blue team finds the difference between the two numbers using jumps on their teams' number line.
6. Both teams use the same recording sheet. The winner of each round is the team with the smallest difference.

Variations

Drawing Two Number Cards 1 through 9

Players use number cards labeled *1* through *9* (Reproducible 1) instead of two dice. Players draw two cards from a paper lunch sack, create two two-digit numbers from the cards drawn, and then return the cards to the sack.

Using Three Dice Instead of Two

Players use three dice instead of two dice. Players roll all three dice and create two three-digit numbers. If two of the three dice rolled have the same number, players must roll the three dice again until all three are different numbers.

Drawing Three Number Cards 1 through 9

Players use number cards labeled *1* through *9* (Reproducible 1) instead of two dice. Players draw three cards from a paper lunch sack, create two, three-digit numbers from the cards drawn, and then return the cards to the sack.

Game Directions

Adding Nines, Tens, and Elevens

Objective

Players take turns recording moves of nine, ten, or eleven along an open number line. Each move is determined by first spinning the game's spinner. The first team to get to the end of the number line wins.

Materials

2 open number lines, labeled *0* and *100*

Adding Nines, Tens, and Elevens Spinner (Reproducible 23)

2 colored markers (1 red and 1 blue)

Players

4

Directions

1. Decide who will be on the red team and who will be on the blue team. Spin the spinner to determine which team goes first (the team with the larger number starts). For the purposes of these directions, the red team will go first.
2. A player from the red team spins the spinner to determine how far to move and then records that move on their team's number line. The red team passes the spinner to the blue team.
3. A player from the red team spins the spinner to determine how far to move and then records that move on their team's number line. The blue team passes the spinner back to the blue team.
4. Another player from the red team repeats the steps, adding onto the existing moves on the number line. The blue team also repeats this process.
5. Play continues until one team reaches or goes beyond one hundred on their number line. This team wins.

Variation

Players use a red and yellow open number path instead of an open number line. Players make their moves by pinning a clothespin or moving a marker along the number path.

Game Directions

Get to the Target (Version 1)

Objective

Players alternate turns, spinning a spinner twice to create a two-digit number (the target number). Players then try to find ways to get to the target number by moving forward or backward using only four types of jumps—twenty-five, tens, fives and/or ones—along an open number line labeled *0* and *60*. The player with the fewest jumps to the target number is the winner.

Materials

1 large paper clip

pencil, 1 per player

Get to the Target (Version 1) Spinner (Reproducible 24)

Get to the Target (Version 1) Recording Sheet (Reproducible 25)

Players

2

Directions

1. Determine who is Partner A and who is Partner B. Partner A should fill out the top half of the recording sheet and Partner B, the bottom half.
2. Partner A spins the spinner twice, creates a two-digit number, and records that number on the recording sheet. (If the spinner lands on same two numbers, the player needs to spin again.)
3. Partner A then draws an open number line labeled *0* and *60* on the recording sheet and marks the target number on the number line. Partner A records their moves on the number line, jumping forward from *0* or backward from *60*, using only four types of jumps—twenty-five, tens, fives, and/or ones—to get to the target number. Once finished, Partner A passes the spinner to Partner B.
4. Partner B spins the spinner twice, creates a two-digit number, and records that number on the recording sheet. (If the spinner lands on same two numbers, the player needs to spin again).
5. Partner B then draws an open number line labeled *0* and *60* on the recording sheet and marks the target number on the number line. Partner B records their moves on the number line, jumping forward from *0* or backward from *60*, using only four types of jumps—twenty-five, tens, fives, and/or ones—to get to the target number.
6. Both players look at each other's number lines to determine who made the fewest jumps. That player is the winner.

Game Directions

Get to the Target (Version 2)

Objective

Players alternate turns, spinning a spinner twice to create a two-digit number (the target number). Players then try to find ways to get to the target number by moving forward or backward using only four types of jumps—fifty, tens, five, and/or ones—along an open number line labeled *0* and *100*. The player with the fewest jumps to the target number is the winner.

Materials

1 large paper clip

pencil, 1 per player

Get to the Target (Version 2) Spinner (Reproducible 27)

Get to the Target (Version 2) Recording Sheet (Reproducible 28)

Players

2

Directions

1. Determine who is Partner A and who is Partner B. Partner A should fill out the top half of the recording sheet and Partner B, the bottom half.
2. Partner A spins the spinner twice, creates a two-digit number, and records that number on the recording sheet. (If the spinner lands on same two numbers, the player needs to spin again).
3. Partner A then draws an open number line labeled *0* and *100* on the recording sheet and marks the target number on the number line. Partner A records their moves on the number line, jumping forward from *0* or backward from *100*, using only four types of jumps—fifty, tens, fives, and/or ones—to get to the target number. Once finished, Partner A passes the spinner to Partner B.
4. Partner B spins the spinner twice, creates a two-digit number, and records that number on the recording sheet. (If the spinner lands on same two numbers, the player needs to spin again).
5. Partner B then draws an open number line labeled *0* and *100* on the recording sheet and marks the target number on the number line. Partner B records their moves on the number line, jumping forward from *0* or backward from *100*, using only four types of jumps—fifty, tens, fives, and/or ones—to get to the target number.
6. Both players look at each other's number lines to determine who made the fewest jumps. That player is the winner.

Game Directions

Get to the Target (Version 3)

Objective

Players alternate turns, spinning a spinner three times to create a three-digit number (the target number). Players then try to find ways to get to the target number by moving forward or backward using only six types of jumps—five hundred, one hundred, fifty, ten, five, and/or one—along an open number line labeled *0* and *1,000*. The player with the fewest jumps to the target number is the winner.

Materials

1 large paper clip

pencil, 1 per player

Get to the Target (Version 3) Spinner (Reproducible 29)

Get to the Target (Version 3) Recording Sheet (Reproducible 30)

Players

2

Directions

1. Determine who is Partner A and who is Partner B. Partner A should fill out the top half of the recording sheet and Partner B, the bottom half.
2. Partner A spins the spinner three times, creates a three-digit number, and records that number on the recording sheet. (If the spinner lands on same numbers, the player needs to spin again.)
3. Partner A then draws an open number line labeled *0* and *1,000* on the recording sheet and marks the target number on the number line. Partner A records their moves on the number line, jumping forward from *0* or backward from *1,000*, using only six types of jumps—five hundred, one hundred, fifty, ten, five, and/or one—to get to the target number. Once finished, Partner A passes the spinner to Partner B.
4. Partner B spins the spinner three times, creates a three-digit number, and records that number on the recording sheet. (If the spinner lands on same numbers, the player needs to spin again.)
5. Partner B then draws an open number line labeled *0* and *1,000* on the recording sheet and marks the target number on the number line. Partner B records their moves on the number line, jumping forward from *0* or backward from *1,000*, using only six types of jumps—five hundred, one hundred, fifty, ten, five, and/or one—to get to the target number.
6. Both players look at each other's number lines to determine who made the fewest jumps. That player is the winner.

Game Directions

Race to 1,000

Objective

Teams take turns drawing a *Race to 1,000* action card. The direction on the action card tells the players how far to move on the number line. After each move, the team then draws a *Race to 1,000* question card and answers the question on the card. The first team to get to one thousand or beyond wins.

Materials

open number lines labeled *0* and *1,000*, 1 marked *Red Team* and 1 marked *Blue Team*, created on a blank piece of paper

2 colored markers (1 red and 1 blue)

Race to 1,000 Action Cards (Reproducible 31), 1 set (made from 1 copy of the reproducible)

Race to 1,000 Question Cards (Reproducible 32), 1 set (made from 1 copy of the reproducible)

2 paper lunch sacks, 1 labeled *Action Cards* and 1 labeled *Question Cards* (to hold the cards)

Race to 1,000 Recording Sheet (Reproducible 33), 1 copy per player

Players

4

Directions

1. Determine who is the red team and who is the blue team. Each team draws an action card to determine which team goes first (the team with the larger number starts; if a team gets the a minus action card they choose again). For the purposes of these directions, the red team will go first.
2. A player from the red team draws an action card, records the move on their team's number line, and returns the card to the paper lunch sack.
3. The other player on the red team draws a question card, answers the question, and returns the card to the paper lunch sack. Both players record the round on their recording sheets. The red team passes the paper lunch sacks to the blue team.
4. A player from the blue team draws an action card, records the move on their team's number line, and returns the card to the paper lunch sack.
5. The other player on the blue team draws a question card, answers the question, and returns the card to the paper lunch sack. Both players record the round on their recording sheets. The blue team passes the paper lunch sacks back to the red team.
6. The red team repeats the steps, adding more jumps onto their number line. The blue team also repeats this process. If a team draws a minus card that puts them backward before zero, they need to place it back in the card pile and draw again.
7. The first team to make jumps to or beyond one thousand on the number line wins.

References

Carpenter, T., E. Fennema, and M. Franke. 1994. *Cognitively Guided Instruction: Children's Thinking about Whole Numbers*. Madison, WI: Wisconsin Center for Education Research, School of Education, University of Wisconsin-Madison.

Chapin, S., and A. Johnson. 2006. *Math Matters, Understanding the Math You Teach*. 2d ed. Sausalito, CA: Math Solutions.

Dehaene, S. 1997. *The Number Sense: How the Mind Creates Mathematics*. Oxford: Oxford University.

Jacobs, V., and R. Ambrose. 2008. "Making the Most of Story Problems." *Teaching Children Mathematics*. 15(5): 262.